NORTH CAROLINA
STATE BOARD OF COMMUNITY COLLEGES
LIBRARIES
ASHEVILLE-BUNCOMBE TECHNICAL COLLEGE

Y0-CUD-047

DISCARDED

JUN 2 4 2025

HOW TO MAKE YOUR OWN SOLAR ELECTRICITY

Acknowledgement

To Leroy Johnson for his inspiration and encouragement. To my wife for all her help and patience. To Virginia Samuels for her 4 a.m. dedication. To Donald Richter and Loana Barlow for their hours of proofreading, and to all others whose help was very much appreciated.

"Since the dawn of man, we have sought to take advantage of the resources around us. We have captured the wind in our sails. We have harnessed the seas. We need only now to harness the sun."
                                                                    Author Unknown

No. 1178
$9.95

HOW TO MAKE YOUR OWN SOLAR ELECTRICITY
BY JOHN W. STEWART

TAB TAB BOOKS Inc.
BLUE RIDGE SUMMIT, PA. 17214

FIRST EDITION

FIRST PRINTING—SEPTEMBER 1979
SECOND PRINTING—MARCH 1980
THIRD PRINTING—OCTOBER 1980

Copyright © 1979 by TAB BOOKS, Inc.

Printed in the United States of America

Reproduction or publication of the content in any manner, without express permission of the publisher, is prohibited. No liability is assumed with respect to the use of the information herein.

Library of Congress Cataloging in Publication Data

Stewart, John W. 1945-
    How to make your own solar electricity.

    Includes index.
    1. Photovoltaic power generation. 2. Solar batteries. I. Title.
TK2960.S74   621.47′5   79-9460
ISBN 0-8306-9747-0
ISBN 0-8306-1178-9 pbk.

Cover photos courtesy of Motorola Inc., Solarex Corp., and SES, Inc.

Preface

Since his early existence on this planet earth, man has always sought simpler and better ways of accomplishing that which he needed done. It seems man has been keenly interested in energy throughout all the centuries of time, but has become extremely dependent on it in the last few hundred years. At first he employed his own muscle power and sometimes that of his friends. History tells how man has at various times used slaves to do much of his work.

Man learned to use and harness the great energy released by burning wood. Wood was used as one of his early fuels to keep him warm and cook his food. When Marco Polo went to China, he found the Chinese burning a strange black rock which he later introduced to Europe as coal. After many years, coal began to replace wood as man's prime fuel source; then, with the invention of the steam engine in the mid-1800's, the Industrial Revolution was born. The birth of the Industrial Revolution began the greatest exploitation of energy sources in the history of mankind, a practice which has continued at an ever-increasing rate until the present time.

With little concern for the future, we have used the richly-stored energy sources of the earth. We have extracted the earth's natural resources at such an alarming rate we are now faced with complete exhaustion of our presently used sources of energy. It seems that man has no appreciation of what he enjoys until it is no longer available to him. In 1972, this was brought to the public mind with the Arab oil embargo, when people were lined up for miles to purchase gasoline for their cars. The average person was stunned by the realization of our growing dependence on other nations for energy. Faced with this dilemma, man is once more seeking new ways to solve his energy needs.

The logical place to look for relief is to the sun, the source of endless energy. With renewed determination, man has gone to work to develop practical and useful solar devices. He has made ovens to cook his food and heaters to heat water and homes. He has even developed air conditioners run by the sun! In his attempt to solve his energy needs, man has once again discovered the extraordinary power of the sun.

In recent years, we have become dependent on electrical power as a convenient form of energy. Great interest is being exhibited in areas where solar energy could be used to produce electricity for use in our homes and businesses. Research in this field has brought to light many ways in which this could be accomplished. Some enthusiasts have proposed using large tracking mirrors to direct the sun's heat on a given point; this would make steam to drive a conventional electric generating system. This requires complicated equipment which is expensive and costly to maintain. One of the most intriguing concepts being developed is a method of direct conversion of the sun's energy into electricity. This process, known as *photovoltaics*, requires no machinery with moving parts to wear out, and if protected, the cells should last indefinitely. The cells can harvest the sun efficiently with no noise or pollution. Photovoltaic cells have been used very successfully in the space program and in many small remote applications on earth.

It is to this subject, its development, uses, and future that this book is directed. Some interesting facts about the sun and man's attempt to harness its energy will be presented, after which the discovery and development of the solar cell will be discussed. We will explain how the cells are made and why they work. Uses of the solar cells now and in the future will be examined. Solar electric generators from simple cells to complete systems will be covered.

It is hoped that through this book, the reader might learn more about the solar cell, how it converts sunlight into electricity, and how it may be used in solving many of our energy needs.

John W. Stewart

Contents

1 The Sun is Up 9
Source of All Energy—The Forgotten Sun—Present-Day Energy Demands—Alternative Natural Energy

2 Electrical Power from the Sun 19
Discovery of Photovoltaic Cells—How Silicon Cells are Made—How the Cell Works

3 Solar Electric Generators 37
The Cell—Large Solar Arrays—Storage Systems—Battery Voltage Regulators—Putting It All Together

4 Efficiencies and Economics of Photovoltaics 57
Solar-Cell Efficiency—High Cost of Cells—Future Goals

5 Applications of Photovoltaics 67
Small Scale Applications—Large Scale Applications

6 Designing a Photovoltaic System 91
Collecting Design Data—Operating Temperatures

7 Cell Materials and Their Development 109
Space Cells—Terrestrial Cells—Photosensors—Current Technology—Improved Manufacturing Techniques—Other Methods of Electrical Conversion

8 Concentrator Cells 127
The Characteristics of Concentrators—Tracking Devices—Types of Concentrators—Concentrator By-Products

9 The Future of Photovoltaics 143
Government Involvement—Looking Ahead—Summary

Glossary 154

Appendix 162

Additional Reading 164

Index 165

The Sun is Up

Probably nothing in our universe has aroused man's curiosity more than the sun. He has studied it, used it and looked to it as the giver of heat, light and even life itself. Without the sun, all life as we know it would cease. Our earth revolves around the sun, the center of our solar system. The sun has always been a scientific curiosity to mankind. He has looked to the sun with great awe and mystery. Seeing it, yet not understanding it, he has long been the recipient of its benefits.

The sun has provided us with light, warmth, food and clothing through endless centuries. It causes the winds to blow, creating a comfortable climate for life. It has purified our water through an evaporative process which has made the rain fall, refreshing the earth.

The very food we eat grew because of nourishment from the sun. Even our own bodies, if shut in without the healthful rays of sunlight, would soon become diseased. The absence of the sun's energy would cause the very earth to become barren and unproductive. Energy from the sun interacts with oxygen in the earth's atmosphere to create a layer of ozone that protects us from lethal doses of the sun's radiation. These are only a few of the ways we directly benefit from the sun's energy.

We also benefit indirectly from the sun, receiving its energy "secondhand." Nearly all of the energy available to us is or was created by the sun. From the sun we receive almost all other forms of power currently available for man's use. The electricity generated by a windmill, for example, is the sun's energy once removed; uneven heating of our atmosphere by the sun causes the wind to blow, thus driving windmills and other useful machines. Hydroelectric generation would not be possible if the sun did not evaporate moisture to replenish our water supplies. The sun's heat evaporates water, causing rain to fall at higher elevations. Water power is created as it rushes down to the ocean. Coal, oil and natural gas were made from plants that required the sun's energy to grow. Energy absorbed by plants and trees provide man and animals with food and energy.

When we contemplate how dependent we are upon the sun, it is little wonder that ancient man at various times worshipped this mysterious object in the sky. The power of the sun and its influence in his daily life made it a natural object for early man to revere. Recorded history cites examples of people who worshipped the sun as their god or one of their gods. Centuries ago, men made temples to their sungod at Stonehenge, England. The sun god Ra is pictured in many of the Egyptian pyramids. From the source of this homage, Sunday, the day so many spend in religious worship, received its name.

SOURCE OF ALL ENERGY

The sun, called by scientists an *orange dwarf*, is 864,000 miles in diameter and weighs as much as 330,000 times the weight of the earth. This fiery ball of gas is composed largely of hydrogen, which is continually being converted into helium by a fusion process. The surface of the sun is called the *photosphere*, a name made from the Greek words meaning light and ball. At the surface, the temperature is 10,000 degrees Fahrenheit, while the inside temperature may reach as high as 300 million degrees Fahrenheit.

The earth, 93 million miles from the sun, is so small and far away that it receives less than one thousandth of one millionth part of the sun's total energy. Even so, this small amount of radiation is equal to 85 trillion kilowatts of energy falling upon the earth constantly.

Every day the sun showers the earth with several thousand times as much electrical energy as we consume. Of this energy, thirty percent is reflected back into space by our atmosphere. Even though only a fraction of the sun's energy reaches the earth, it provides us with all the energy we can possibly use. The amount of sunlight striking the earth every fifteen minutes could be enough to satisfy our total energy requirements for an entire year. The solar energy reaching the earth three days is greater than the estimated total energy of all the fossil fuels on earth. The amount of solar energy received each year by the United States alone is equal to 1150 billion tons of coal. In fact, the average solar energy striking the roof of a typical residence is 10 times as great as its annual heat demand.

Why are we not making use of this tremendous source of energy provided by the sun? The answer, of course, is that we have been using it for some time. However, instead of using the sun's power directly, we have been using stored or indirect solar energy in the form of fossil fuels. These were created by the sun and trapped in the earth eons ago, and only recently reclaimed by drilling and mining. But we must now begin using this power directly from the sun. It is estimated that the sun will be shining brightly for at least another five billion years, so solar power is considered inexhaustible. This ever-present, furnace-like sun offers man his only free, universally available source of power which continues to fall on the earth, waiting only for man to reach out and utilize it. Because of the tremendous potential of energy both for the present and future, we must look to the sun to provide our future energy needs. Consequently, solar energy could be providing substantial levels to fill the public's need within the next 10 years.

THE FORGOTTEN SUN

From the dawn of time, man has found sources of energy to help him do his work. He used wood for many years to warm his home and cook his food. The discovery of coal made it possible to use this rock to satisfy his needs. And with coal, he began to develop even greater power sources to work for him. With the widespread use of coal and the invention of the steam engine, the Industrial Revolution was born. The steam engine made the railroads possible. Men like Tesla and Edison soon developed electrical generating plants to be driven by steam. The internal combustion engine was developed along with the advent of gas and oil. The automobile industry soon adapted this engine to such an extent that all others seemed to disappear. Fossil fuels provided power for all these new inventions. Because industrial development was dependent on fossil fuels, the costs associated with using stored solar energy became significantly less than the cost of developing means to collect and utilize energy received directly from the sun.

History is full of examples of man's trying to harness the power of the sun. A Greek scientist, Hero, made wind- and sun-powered devices. In 212 B.C., Archimedes set fire to Roman ships at Syracuse using polished shields. Water wheels and windmills were soon capturing an indirect solar power. For centuries, sundials were used as a way of telling time. Sailing ships made it possible for man to discover the far away lands of the earth. The clipper ship *Sovereign of the Seas* traveled 485 miles in a single day. Between the 14th and 17th centuries A.D., many solar devices were invented. The solar engine invented by Salomon de Caus of France used sun-heated air to run a water pump.

In the 18th century, a solar furnace designed by a French scientist, Antoine Lavoisier attained temperatures of 1750° C or 3182° F. In the 19th century, the famous Stirling engine was developed, which can operate by the sun. This engine ran everything from printing presses to water pumps. Up to this time, most of the work with the sun was with devices that focused the sun's rays on a single point. In the late 19th century, a new device appeared on the solar horizon: the flat-plate collector. A flat-plate collector uses the sun as it falls upon its surface without concentrating the rays. This procedure proved to be a less expensive way to harness the sun. The next 50 years brought many different devices of this kind into play, all using the sun to work for man.

With the discovery of coal and later oil, however, solar-powered gadgets made way for these less expensive forms of readily obtainable energy. The vast reserves of stored energy were now being used to supply energy to man. Because these fuels were already in a form that could be easily obtained, transported and stored, and required no elaborate method of collecting, fossil fuels were preferred to the sun. However, as world fossil fuel demand has increased and conventional fuel supplies have decreased, the direct use of the sun's energy has emerged as a powerful alternative to man's increasing energy and environmental problems. The advantages of solar energy are already becoming readily apparent. Areas such as cooking, heating, refrigeration and light temperature steam generation are already successfully used and developed. Additionally, using the sun to desalinate sea water could solve some of our world's water shortages.

PRESENT-DAY ENERGY DEMANDS

Of all the energy used in the world today, over 96 percent is from fossil fuels. Because there has been an abundance of fuel, such as nature's wood, coal, oil and gas, people have long viewed these resources as their birthright. These sources of energy have powered our automobiles, heated our homes, and fueled our industries. Where these fuels were once abundant and low-priced, they are now becoming scarce and expensive.

Due to the ease of electrical distribution, the demand for its convenience, and its use in lessening pollution, electrical production will require the highest future increase. The total energy used in the United States for electrical power generation will increase from 25 percent in 1970 to 36 percent by 1985. From 1950 to 1970, electric consumption increased annually at the rate of 7.5 percent. This increase in consumption is expected to remain relatively constant through 1985.

Most of the electricity used in the United States comes from changing water into steam that in turn drives some type of turbine. Coal, oil or some other type of fossil fuel is used to produce the steam for these generating plants. To see how we consume our energy, let's look at the sources that now supply our electricity.

Hydroelectric power contributes only 4.1 percent of America's energy and has been used to its maximum. At the present time, we have used practically all of the hydroelectric sites that are available in the United States and other parts of the world. Few locations for dams remain which could be used to produce electrical power, and a number of those will probably be eliminated after environmental impact studies are made.

Nuclear power produces only 1 percent of our nation's electrical power. Because the use of nuclear reactors is relatively new and has produced problems in waste disposal, it has met with poor public acceptance. But even nuclear power has a limited source of fuel available. Because of these problems, it doesn't appear that the use of nuclear power will have a substantial impact on our current and future energy needs.

Coal, now providing approximately 20 percent of our energy, is being revived as a major source, but this, too, is finite. If we shift too heavily in this direction, our coal reserves will soon be depleted.

Natural gas, providing 32 precent of our total energy needs, is becoming difficult to obtain. In many states, new natural gas hookups are no longer available due to shortages.

Oil, which we are now using more than any other source, supplies 44 percent of our current energy requirements. If we continue to use oil at this high rate of consumption, it, too, will soon dwindle. Today we are so reliant on oil and natural gas to heat our homes and factories that we would be in serious difficulty if we could no longer get these precious commodities.

It appears that only through the jolt of the Arab oil embargo were we awakened to the realization of our growing dependence on fossil fuels. We had been going merrily on our way, spending the money from our energy bank as though it would last forever. Suddenly we received our overdraft notice. We then had to borrow from someone else to cover our deficit. Because of this, we are now getting almost 30 percent of our energy from foreign sources, and this figure increases daily.

Instead of using and developing other sources of energy, Americans have practically exhausted our known sources of oil and natural gas. Our domestic supplies are dwindling, forcing us to import more fuel at higher prices. In 1972, the United States produced about 87 percent of its energy requirements. With the increase in energy usage and our dwindling supply, our dependence on foreign sources of fuel is expected to increase dramatically. Although it has only 6 percent of the world's population, the United States consumes almost one-third of the total energy in the world each year.

In his attempt to keep warm and find energy to do his work, man has indiscriminately used fossil fuels to such an extent that they are fast disappearing. Predictions are that in less than 100 years, fossil fuels will be depleted. Besides the fact that we have used up our sun-created oil reserves at an astronomical rate, we have also played havoc with our environment. We begin to see the pollution we have created through our indiscriminate use of our natural energy resources. Exhaust from burning fossil fuels from our cars, homes and factories has polluted our environment to such an extent that we can hardly breathe or see the world around us. Our eyes burn from the smog in the skies, the fish die in our polluted rivers and lakes, and we are being poisoned by the food we eat and the water we drink. The elimination of wastes has destroyed and polluted our surroundings. The strip mining of coal has ravished the landscape. The wastes from nuclear power plants threaten our very existence. Many of the once-beautiful beaches are now covered with oil.

It seems to be the nature of men that when we find something which performs well for us, we begin to exploit it. We have abused the earth for our own gain. In a short 50 to 60 years, we have used up most of these abundant reserves of energy. Only now do we realize that these unreplenishable sun-created stored sources of energy cannot last forever. We have started a trend which will not easily be reversed. Yes, in a brief space of time, we have turned our world upside down, only to realize we have destroyed the beauty and tranquility of our once uncomplicated existence.

With the sun shining brightly overhead, we are now experiencing an energy crisis. Faced with the problems of an environment polluted through by-products of our technological society, and with the rapidly-dwindling supplies of presently used sources of energy, we are now looking at alternative ways to fill our energy needs. The need for economically attractive, nonpolluting alternative energy sources of power other than those presently being used if our future activities continue at their ever-increasing rate.

ALTERNATIVE NATURAL ENERGY

For years, man has used *wind power*, which has served him well. Sailors have used the wind for ages to power their ships. Man has used the wind to grind his wheat, pump his water, and is now considering it an an alternate source of electricity. Perhaps giant windmills will someday be able to provide electrical power for cities. The major difficulties with using the wind to generate electricity are that the wind doesn't blow all the time and that practical storage systems still need to be developed.

For centuries, *wood* has supplied us with heat for our homes and with fuel for cooking. It was one of the earliest forms of energy used by man. Wood is in such high demand today that many are concerned about whether we will have enough to last in the years to come. Because of the shortage of wood, some people are growing trees on farms to provide wood for paper and fuel. Wood has served us well as one of the major building materials for our homes and is now being reexamined as an alternate energy source to heat our homes. Here also, we must be careful not to overuse this precious natural resource or it, too, will soon be gone.

Men are now working with ways of harnessing the power of the *ocean waves*. By trapping the water at high tides and allowing it to run through a generator as it flows back to the ocean level, considerable power could be provided. The problem here is finding a way to harness ocean power economically and to distribute it from where it is obtained to where it is needed. The up and down movement of the waves could also offer a great source of power.

Many are looking to *geothermal* energy, the hot water and steam deep inside the earth, to generate electrical power. Scientists believe that the center of the earth is made of a molten mass. In some areas, this mass is close to the surface, evidenced by volcanos, geysers and hot springs. If this heat could be trapped and used to generate electricity, it could supply much of our electrical need without polluting the atmosphere. The problem is that there are only a limited number of places where the energy can be tapped because generally it is too far from the earth's surface. Considerable time and expense would be required to recover the heat. It is not expected that geothermal power will have a significant impact on the total energy picture before the year 2000.

The percentage of help from *hydroelectric* power is expected to decrease in the future instead of increase. There may be some effort to use off-peak electrical power to pump water back up to the storage pond to be reused during peak electrical requirements.

Much attention has been given to the process of controlled *nuclear* fusion, but perfection of this process is still many years away because of technological and research difficulties.

The conversion of waste heat into electricity, commonly termed *thermal electrics*, could be a viable source if harnessed properly. The problem is in collecting the waste materials to be burned in an economically feasible manner.

It is possible to make a burnable gas, methane, by heating decaying waste products in an oxygen-free environment. Many of the wastes that could be used in this process are now polluting our surroundings. The difficulty with this process is also in gathering waste into a processing plant.

Among the many energy alternatives being examined, *solar* energy ranks high as a consideration. Of all the alternate sources of energy, the sun has the greatest potential as a universal form of power (see Fig. 1-1).

Harnessing the sun's power is considered an attractive alternative because it is a renewable resource which causes no pollution. In contrast to conventional fuels, its use eliminates the need for refining, transporting and conveying fuel and power over long distances. The use of solar energy for heating and cooling promises a more rapid payoff than other energy alternatives because the basic technology already exists and needs only minor refinements. Considerable research, development and demonstration activities have been initiated in the public and private sectors to facilitate the widespread utilization of solar energy.

Although solar energy is available and free, its capture and utilization is not. There are some very difficult technical, social, and economic problems which must be resolved in order to change the existing patterns of energy conversion and consumption. The problem now is to economically convert the sun's energy into a useful product which can be consumed by man.

Through the wise use of solar energy in our homes and industries we may begin to retard the trend of using up our supplies of oil and gas. The widespread use of solar energy could reduce our rapidly-growing dependence on these forms of energy, which are polluting our surroundings and daily rising in cost.

Solar energy is not centralized but is available to all. It can be collected by each person at his own home, thus reducing distribution

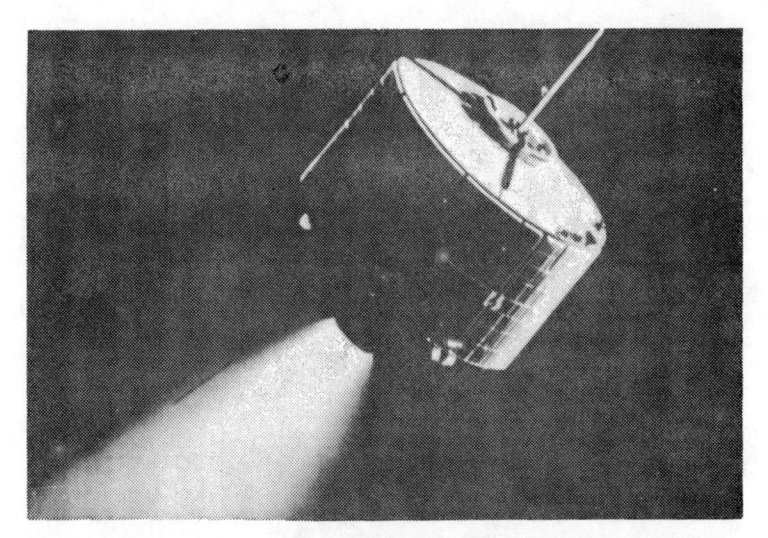

Fig. 1-1. A man-made space satellite covered with electricity-generating solar cells. The space program used photovoltaic cells first (courtesy Solarex).

and transportation costs. It would not be subject to governmental controls, business monopolies, or foreign boycotts. Solar energy is a free gift to all regardless of color, creed, or political involvement. It is distributed equally to the rich and poor. Yes, a bright new age of solar energy is dawning.

Electrical Power from the Sun

2

In recent years, the demand for electrical energy has increased dramatically. Between the years 1970 and 1976, the United States' total energy consumption increased some 4 percent. According to the *National Petroleum Council's Report on the United States Energy Outlook*, about 22 percent of the total energy consumed is used in residential and commercial heating and cooling, 30 percent in industry, 24 percent in transportation, 6 percent in nonenergy such as plastics, fertilizers, etc. and 21 percent in electrical generation. See the pie chart in Fig. 2-1.

Most of the electrical energy generated in the United States comes from oil, coal or natural gas. Of all the energy consumed in the United States, about one-quarter is provided by electricity, a convenient and versatile form of energy. The *United States Energy Outlook* further states that it is expected that the use of electrical power will increase to 33 percent by 1985 and to 50 percent by the year 2000.

With the exception of small contributions of hydroelectric and nuclear power, fossil fuels have been heavily relied upon, providing 90 percent of the electricity being generated nationally. These fuels are burned in a furnace to heat water to the state of steam. This steam is used in conventional turbine engines, driving electrical generator shafts to produce electrical current. This is a very roundabout process with relatively low efficiency. But the need for continued increase in the production of electrical energy seems imminent. This clean form of energy is tremendously popular in heating and lighting our homes and industries. Electricity can be transmitted over wires right to the users' homes. Until now, it has been competitively priced in most areas with other fuel sources. The low efficiency with which electricity is generated by using fossil fuels has made many feel that these fuels should be used for other, more practical purposes.

If the 24 percent efficiencies of electrical systems were reviewed, looking beyond the efficiency ratings of the Btu produced in their burning to the time involved in producing the coal or oil from the sun's energy thousands to millions of years ago, the overall efficiencies would be reduced dramatically. Consider the amount of sunshine it takes to grow the plants and trees, the years involved in making coal from these plant materials, and the small fraction of heat returned by burning them.

The use of fossil fuels to generate electrical power also has some adverse environment effects. Much of our air pollution in large, developed areas comes from the stacks of the generating plants. Huge transmission lines have been erected, marring the landscape from coast to coast. Construction, maintenance and line loss account for much of the present cost of electricity. These lines are subject to wind, snow and lightning, and are vulnerable also to sabotage.

We must now begin to look for more efficient ways of generating our electrical power to save our fossil fuels for other purposes. We can no longer use our fossil fuels at such alarming rates to produce electrical power but need to conserve these limited natural resources. With the increased use of electrical power, experts have proposed using the sun to generate electricity. So as we look for alternate sources of energy, our attentions need to be directed toward some practical means of producing electrical power from the sun. By using the sun in this manner, we will have a much more direct and efficient system, one free from pollution.

Several methods of converting the sun's energies into useable heat have already been developed. Most commonly, heat is collected by letting sunlight strike the back plate of a collector, which absorbs the incoming radiation and transfers it to a fluid of water or air. A well-designed collector may have one to two layers of glass or plastic over the absorber plate to reduce the losses into the atmosphere. Overall effectiveness of 40 percent to 75 percent has been achieved from these types of systems. Losses occur because of inefficient collector design and reflection. These units are now being installed all over the world in such applications as water heaters, pool heaters and even some space heating. Heat for generating electric-

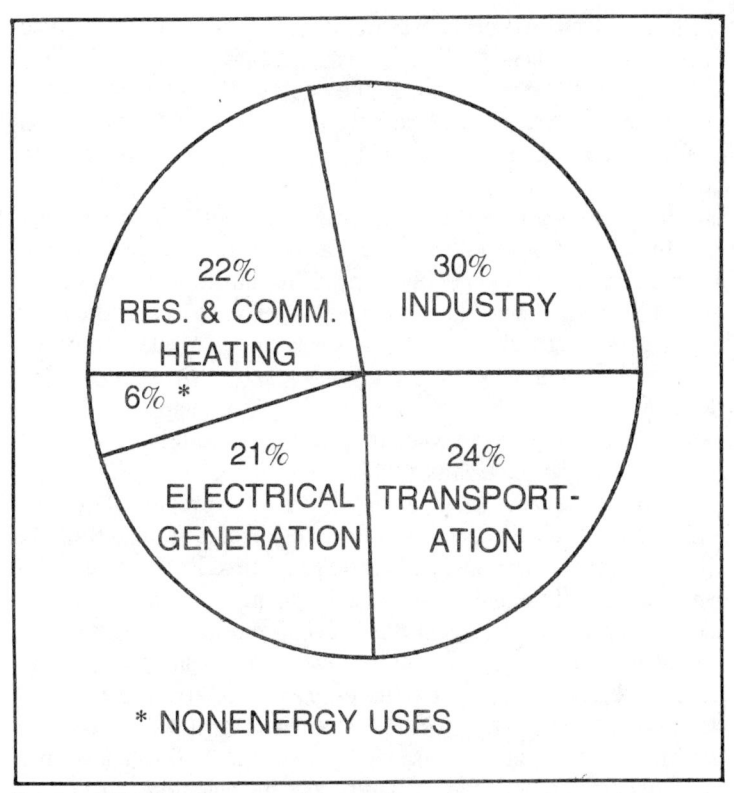

Fig. 2-1. Chart showing energy consumption.

(Pie chart: 22% RES. & COMM. HEATING; 30% INDUSTRY; 6% * ; 21% ELECTRICAL GENERATION; 24% TRANSPORTATION; * NONENERGY USES)

ity could also come from the sun through a process called *solar thermal conversion*. By using concentrator collectors, higher temperatures could be achieved with somewhat lower efficiencies. A collector of this type could provide high temperatures by concentrating the sun's rays with mirrors or lenses onto a metal pipe containing a liquid. This type of system could produce enough steam to run electrical generating turbines. The primary restriction on this type of solar thermal power plant is that it must be used in areas of ample sunshine. In the southwestern United States, where year round sunshine is available, such plants are being constructed. These plants still have all the inherent problems of a conventional steam generating plant and the power must be distributed on a network of utility lines. They also require elaborate tracking devices and are costly to maintain.

There is another way to produce electricity directly from the sun. It is an area which is receiving a great deal of interest and has a very exciting future. The *solar battery*, as it has been called, has been producing electricity for 20 years in outer space (see Fig. 2-2). The device used for such conversion is called a *photovoltaic cell* or *solar cell*. The phenomenon known as the photovoltaic effect allows the conversion of light energy directly from the sun into electrical energy without any moving parts. By using the photovoltaic cells to generate electrical power, the energy-wasting steps involved in the thermal process can be eliminated. Instead of converting sun to heat as in the thermal process, solar cells can be employed to convert the sun's energies directly into electricity. This effect occurs when the energy from the sun hits certain light-sensitive materials, thus creating an electrical current. This small device is able to produce electricity without using the costly generating plants required in the thermal conversion. It uses no large or costly transmission lines to deliver its power, but can be employed at the site. It has no moving parts to wear out or maintain and is simple to install.

Much is needed, however, in the way of improved efficiencies and lower prices for these systems to become a cost-effective alternative to the customary generating facilities. But it seems to be only a matter of time until improved techniques of generating electricity from the sun through this process can be developed. It is relatively certain that this type of generating facility will begin to have increased popularity for the production of electricity and for minimizing the amount of fossil fuels required for this purpose. The use of the sun's energies in producing electricity in this manner has many exciting potentials and would reduce the pollution caused from the conventional oil-fired and coal-fired generating facilities throughout the country.

DISCOVERY OF PHOTOVOLTAIC CELLS

The process of converting light energy from the sun into electricity was first discovered in 1839 by Antoine Bicquerel. This fascinating but difficult to understand process had no intervening steps or complicated collector. It seemed only to be a scientific curiosity at first and many years went by before practical use of the solar cell was applied.

At first the element used in solar cells was selenium, receiving its name from a Greek word for the moon. Selenium had very low efficiencies of converting the sun's energies into electricity. Because of its low efficiency, it was impractical to use cells for any large operations. Gaining popularity in the photovoltaic market, selenium was first used in light meters, photoelectric devices, and the like. Today, because of improved production techniques, selenium cells are relatively inexpensive, but are still quite inefficient.

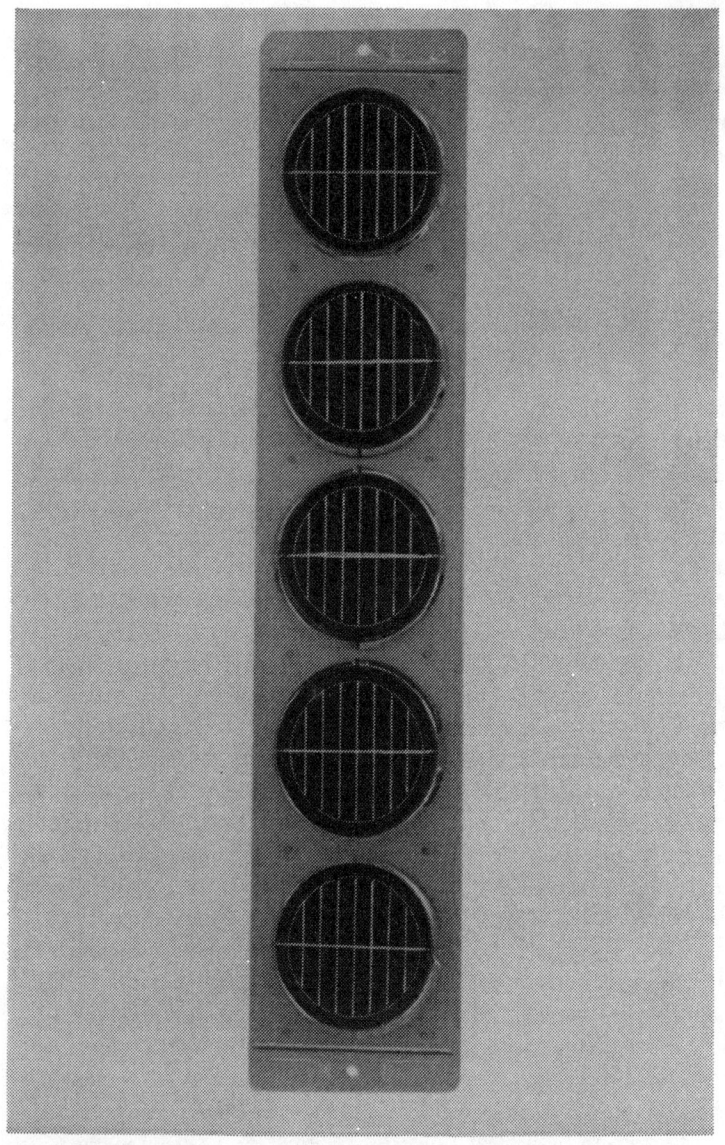

Fig. 2-2. A solar battery consisting of five cells (courtesy Silicon Sensors, Inc.)

In 1873 Willoughby Smith selected a type of material called silicon for experiments in underwater telegraphy and found it had good properties for producing an electrical current when exposed to light. Since that time, many scientists and researchers have devoted much of their time and attention to ways of using silicon to produce electricity.

Silicon is one of the most abundant elements on the earth. It is found in ordinary sand but also appears in compounds that require much refining to produce the pure silicon needed to make the solar cells. Most of the solar cells today are made of silicon, which has proven to be one of the most efficient materials used to date in making photo cells.

The development of solar cells made from silicon is a direct outgrowth of the transistor industry, where experimenters used silicon for making transistors and then gradually switched to germanium. At a later time when efforts were made to develop a more efficient photovoltaic cell, the electrical properties of silicon were remembered. In the 1950s, researchers discovered how to make solar cells using polycrystalline structures of metallurgical silicon. This material was used to develop a more efficient solar cell. Bell Laboratories produced silicon cells with an efficiency of more than 6 percent. This meant that of 100 percent of the sun's energy falling on the cell, 6 percent was turned into a useable form of electrical energy. Even though this seems like a small percentage, it is many times greater than the direct conversion of the sun's energies through the fossil fuel process.

In the 1960s Dr. Joseph Lindmayer of Comstadt Corporation invented the violet cell. Efficiencies increased up to 18 percent as the cells were produced for the space industry, and solar batteries or solar cells became famous through their use in this program. Until 1974, most of the cells being produced in the United States were being used principally in the space program. Solar cells have provided reliable electricity for a variety of space vehicles for some 20 years. Many satellites are now sending signals back to earth through the use of sun-powered transmitters.

The first practical terrestrial solar cells were manufactured to supply small amounts of power to remote weather equipment. Bell Laboratories went on to show how the solar cell could be used to power telephones with sunshine. Today rural telephone lines and remote power stations are proving to be useful and economical applications. A message was sent by a solar-powered transmitter and picked up by a solar-powered receiver in 1955 by some ham radio operators (see Fig. 2-3).

Solar One, a home built by the University of Delaware Institute of Energy Conservation, has a portion of the roof covered with a type of solar cell collectors. The excess heat was carried away from the cells and fed into a eutectic salt storage. The electricity from the cells was delivered from the local utility company. This electricity was used to operate many small electrical appliances and devices.

Fig. 2-3. Solar remote power generator (courtesy Solar Energy Systems).

The solar cell is a remarkable little device that has revolutionized the solar industry. Constant progress has been made since they were first used in photography as light meters. Electricity generated by the solar cells coming directly from the sun is a gift compared to the ways we presently get our electrical power. One of the advantages of photovoltaic cells is that they can be used without power lines. They also offer long life expectancy with no moving parts and extremely low maintenance. Photovoltaic systems offer a highly reliable source of electric power. They are capable of operating efficiently in either small, low-power remote applications or in large central power stations for producing power for the utility companies.

Photovoltaic cells receive energy from an inexhaustible resource that could provide sufficient quantities of electrical energy. There is no need to worry about fuel for photovoltaics as they use the only nondepleting, nonpolluting and ever-present source of energy, the sun. The process of direct conversion of solar radiation into electricity by photovoltaic arrays is clean, silent and involves no heat cycle or moving parts. The solar electric system gives off no toxic by-products, no smoke or any other bothersome waste products. A wide range of applications, types, sizes and shapes is permitted due to the modular aspect of the photovoltaic array. With the high demand for electrical power, photovoltaic conversion could greatly reduce the fossil fuels now being used in the production of

25

electricity. The use of photovoltaic power generation would be an enormous benefit to mankind if it were used to provide him with sufficient amounts of electrical energy.

Although the process of converting light into electricity has been known for many years, the barrier to the large-scale terrestrial conversion of sunlight into electricity has been their high cost. The materials used and the laboriously intensive hand production process make the cells very expensive.

To understand why cells are so costly and require so much labor to produce, let us examine first how the silicon cells are made. After we have a better understanding of how they are made, it will be easier to understand how they can produce electricity. More detailed descriptions of the production of the cells will be discussed in Chapter 3.

HOW SILICON CELLS ARE MADE

Silicon, another name for sand, is found in abundance on the earth's surface. However, the silicon used in making solar cells must be 99.999 percent pure according to several manufacturer's publications. To purify sand to this degree, much refining is required. This high-grade silicon is termed *metallurgical-grade silicon*. It has been used for several years in the transistor industry. The problem with this high-grade silicon is the cost. It may be as common as sand, but pure silicon as used in the photovoltaic process costs about $33 per pound. Yes, the basic element of the solar cell is one of the most abundant natural resources found on earth, but its simplicity seems to end there, for silicon must also be grown into crystals before it can be used

To grow a crystal, the silicon is placed in a crucible and heated to 205,000-plus degrees Fahrenheit. At this temperature, a seed crystal is placed at the end of a rod and lowered into the molten silicon. The molten liquid adheres to the rod and the crystal begins to grow. As the crystal grows, the rod is pulled and twisted in an upwards direction, causing the hardening crystals to form in a definite pattern. These ingots grow very slowly, only a few inches per hour. The crystals now forming in the ingot are all growing in the same direction. This is very important to ensure the proper response of the cell. The diameter of the ingot can be controlled by the rate of pull. Most ingots used for solar cells are about 2 to 3 inches in diameter, but some are larger. See Fig. 2-4 and 2-5.

After cooling, the hardened ingot is sliced into very thin wafers with a diamond saw. Because the saw blade is several times thicker than the cell itself, more than 50 percent of this costly material is

Fig. 2-4. Solar cell at various stages of development (courtesy Motorola).

wasted. After the cells are cut to the desired thickness of about 0.012 inches, they must be polished.

These thin slices of silicon are poor conductors, so they must then undergo a series of steps to change them from an insulator into what is termed a semiconductor. The two surfaces of the cells must be treated with impurities, creating dissimilar electrical properties from side to side. The process of adding these impurities is called *doping*, as shown in Fig. 2-6. The cells are arranged on a rack and inserted into a heated chamber. One side of the cell is exposed to

Fig. 2-5. Hardened ingot (courtesy Motorola).

boron atoms having an extra positive charge in their molecular structure. The other side is exposed to phosphorous or arsenic atoms, having a negative charge because of the missing electron in their atomic structure. This negative charge forms *holes* as they are called, to attract the electrons from the boron atom. As the wafer is heated, these elements are allowed to penetrate the surface of the cell. The amount of penetration is a highly controlled process and must be monitored. As the two sides of the cell are exposed to the desired depth with these different materials, a neutral area is left between the sides. This neutral area is called a PN junction, shown in Fig. 2-7. It receives its name from the positive and negative sides that it separates.

The cell is then exposed to an etching process so that a metal coating can be applied. In order for electrical charges to be harvested, a metallized grid is applied to the surface of the cell. The purpose for this is to harvest as much electricity as possible without shading the cell from the sunlight. Too thick a grid would result in less energy being received from the sun. If spaced too far apart, it would result in not being able to harness the power generated because the electron must travel too far to find the grid. The grid appears like a system of roads all leading to a highway. Many patterns have been designed and employed by various manufacturers. This grid must be applied by a photographic process similar to

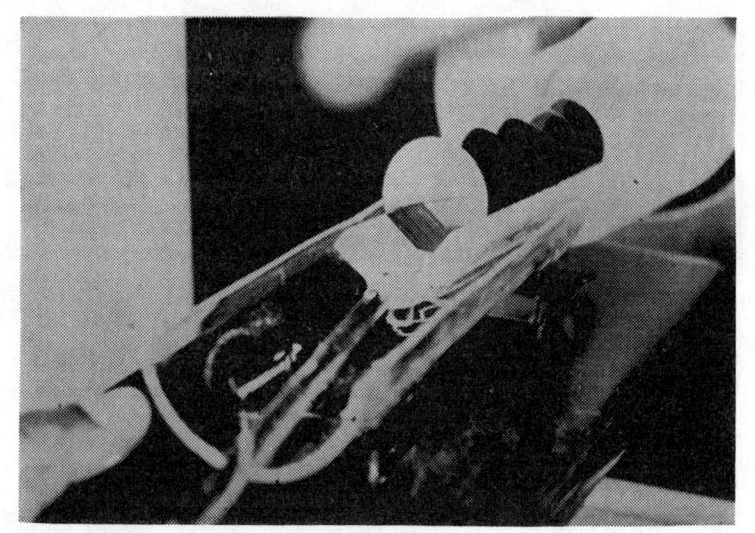

Fig. 2-6. Doping the cells (courtesy Motorola).

silk-screening with the pattern being etched on the cell. A flux solution is applied to the cell to make the metal adhere to it. The cell is then immersed in a solder bath, where the metal adheres to the cells, thus forming the grid. Several different types of metal can be used in applying the grid, but normally they are either gold, silver or aluminum. When the silicon cells are first cut from the ingots, they have a silver glossy look, but as they are doped, they become dark blue. The metal grid is placed on the cell and it takes on its final

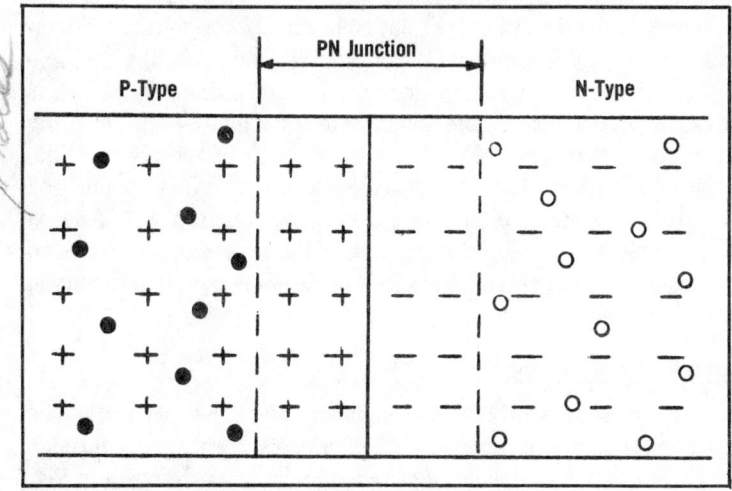

Fig. 2-7. Cross section PN junction.

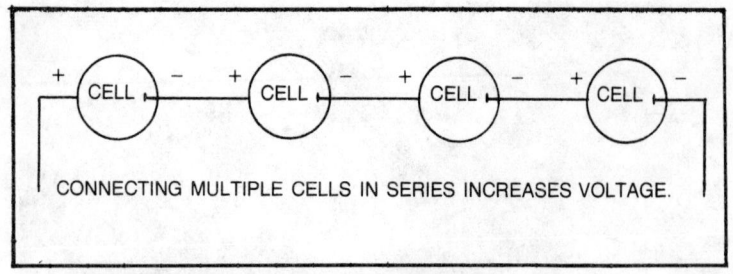

Fig. 2-8. Series connection of solar cells.

appearance. The back of the cell is generally completely coated with the solder bath.

The top side containing the grid is then covered with a special antireflective coating to prevent the sunlight from being reflected off the cell. This increases the amount of sunlight absorbed and improves the overall efficiency of the cell. This antireflective material has a textured surface which greatly enhances the amount of light absorbed by the cell. Some manufacturers apply coatings that reflect 6 percent to 7 percent of the infrared radiation to reduce heat buildup in the cell. After the cell has received the antireflective coatings, it is ready to be tested and inspected to see if it has the desired efficiency and quality. Many cells, which for some reason or another, do not pass the final inspection. Some are broken in handling; in others the solder bath has bridged across from one side to the other causing a short in the cell.

Upon passing this inspection, the cells are packed ready to be shipped to the module assembly line. Many companies prefer to assemble their own modules, but cells can also be purchased individually. However, one cell will not provide much power by itself. So cells are assembled into modules for most applications. A *module* is basically several cells connected in series (see Fig. 2-8) by soldering leads to the trunkline of each cell to the back of another until the desired voltage or amperage is achieved. These cells then must be housed in a protective covering, usually glass on top with a silicon rubber holding the cells in suspension. A rigid frame is placed around the glass to protect it as well as to provide some means of mounting (see Fig. 2-9).

HOW THE CELL WORKS

There is really no simple explanation of how the cells work. The solar cell actually converts visible light into electrical energy through a process called the *photovoltaic effect*, which is illustrated in Fig. 2-10. The solar cell is a semiconducting device made of materials

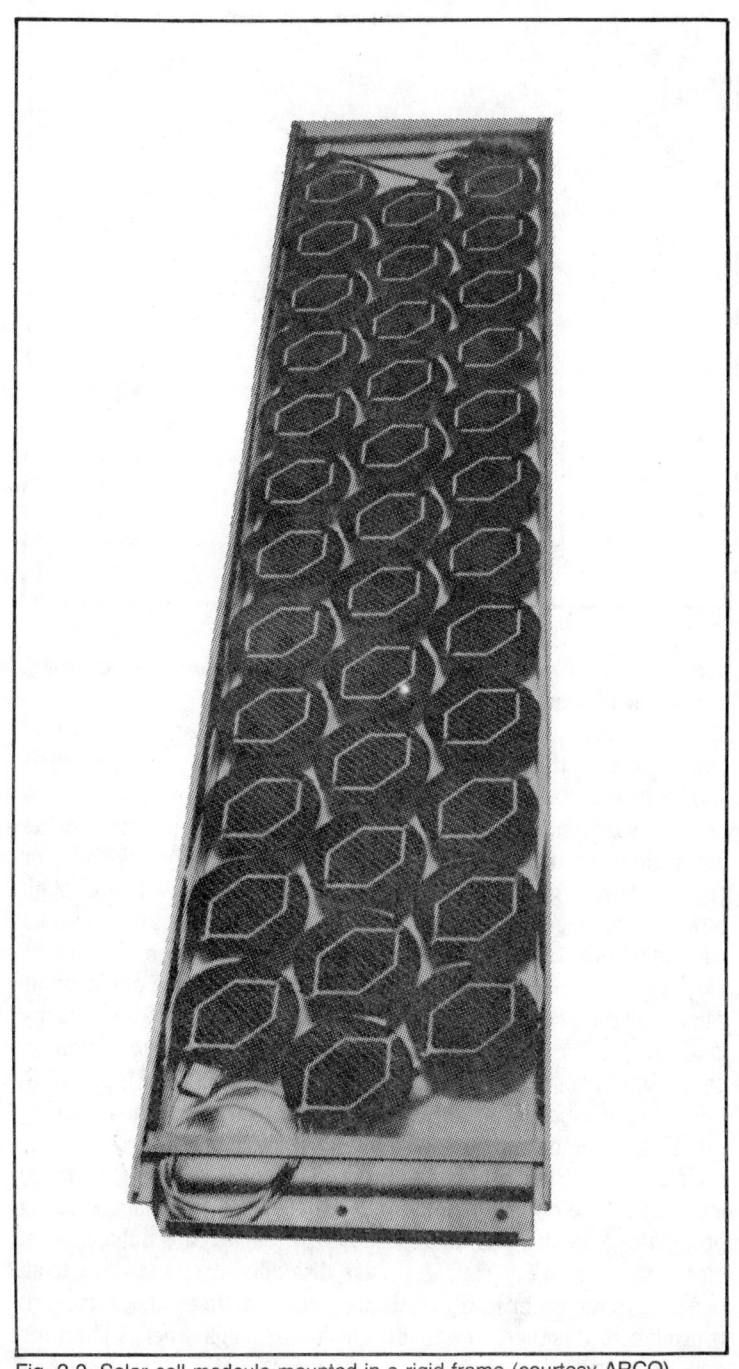

Fig. 2-9. Solar cell modeule mounted in a rigid frame (courtesy ARCO).

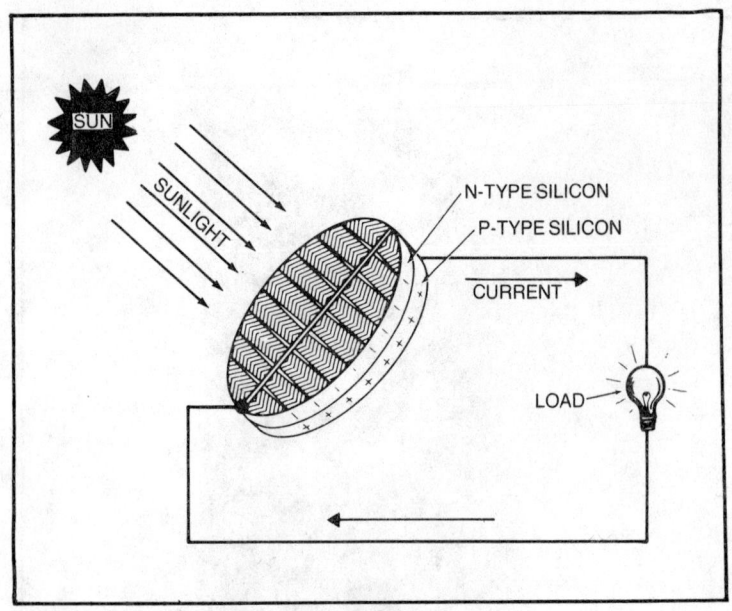

Fig. 2-10. Photovoltaic effect (courtesy Solarex).

which produce electricity when exposed to light, but remain dormant in dark or shaded conditions.

To better understand how the cells function, perhaps a brief explanation of the atomic theory is needed. The silicon atom is shown in Fig. 2-11. The atom consists of a core of neutrons and protons surrounded by revolving electrons. The electrons in the outer shell are the most effected by the light energy. *Photons*, or bundles of energy from the light, strike the cell providing enough power to dislodge the electrons. When one side has extra electrons and the other is missing them, there occurs an attraction of the extra electrons to the side missing electrons. The imbalance of electrons caused by the dopant results in movement of electrons when energy is applied to the cell. This imbalance remains until acted upon by some source of light bright enough to set them into motion. In designing solar cells, materials that perform best are those having the greatest imbalance in the number of electrons in orbit around their nucleus. Silicon is basically an insulator which prohibits this attraction. The doping process gives the silicon its semiconducting properties. As the cell is exposed to the light, the imbalance of electrons tries to equalize. As these dislodged electrons flow from one side to the other to try to create a balance, an electrical current is produced. It is theorized that each photon of light energy liberates one electron.

The light shining on the upper side of the cell produces an excited condition in the electron band of the molecular structure. As sunlight strikes the surface of the cell, negative charges, called *electrons*, and positive charges, called *holes*, are set free. As the electrons become more excited, enough power is produced to cause an interchange between the two opposite sides producing a voltage. Because of the crystalline structure of the silicon cell, the electrons are forced along from one atom to the other. An insulator wall exists between the two opposite charges causing the current to flow around through the wires and load to the opposite side. By placing leads on the opposite sides of the cell, the electrons will flow from the negative to the positive side, completing an electrical circuit.

The metallic coating placed on the back of the cell and the vein-like network etched into the front of the cell are to receive the voltage.

A very small amount of current is obtained from a single cell, but when cells are hooked together in series, a considerable amount of power can be obtained. The electrical current produced is approximately 0.45 volts from any cell no matter how large or small it may be. Although the amount of current varies with the surface area of

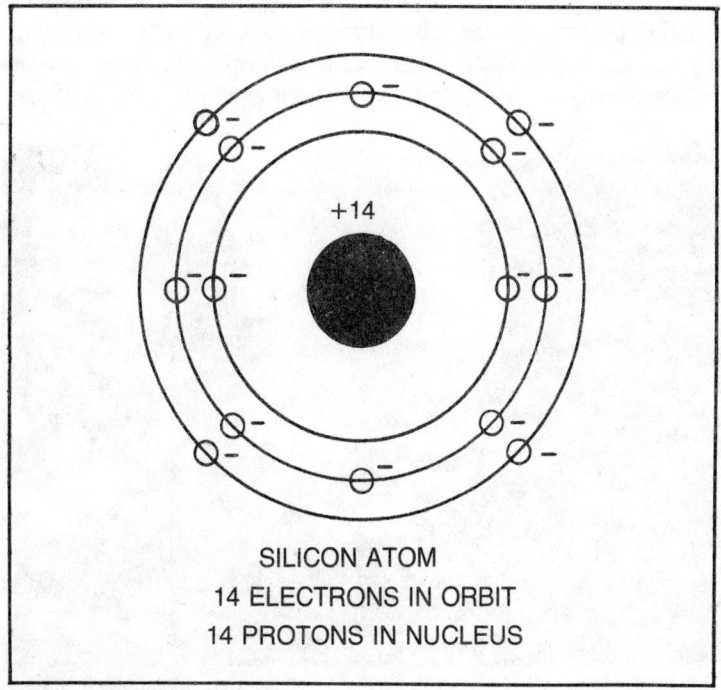

SILICON ATOM
14 ELECTRONS IN ORBIT
14 PROTONS IN NUCLEUS

Fig. 2-11. Atomic theory.

the cell and with the intensity of the light, the cells respond to a broad range of light, from ultraviolet on one side of the spectrum to infrared on the other. Electricity is thus produced whenever there is light. Some light from the sun strikes the surface of the cells even on overcast days. The cells produce current under artificial lights, either incandescent or fluorescent. Fig. 2-12 illustrates a demonstration using artificial lights. The more intense the light, the more electricity received from the cell. And thus, as the sun goes behind clouds, there is a considerable reduction in the amount of electricity which can be obtained. These cells are very sensitive to the amount of light falling on them. The amount of current or power delivered by the cell is dependent on its size and efficiency. The cells are connected electrically into solar modules, the building blocks of the solar electric system. As the basic building blocks, the modules can be connected together in parallel or in series to create any amount of voltage or amperare needed to fit the desired application.

Because the function of the solar cell is similar to the operation of a battery, many people have termed it a solar battery. The one important difference between the solar cell and a dry-cell battery is that the solar cell, if protected from damage, will continue to produce indefinitely, whereas the dry-cell battery will eventually wear out. Another difference is that a battery can be used to store energy, but a solar cell only produces while exposed to light. However, a solar cell can charge a battery with the power it generates. The fact that

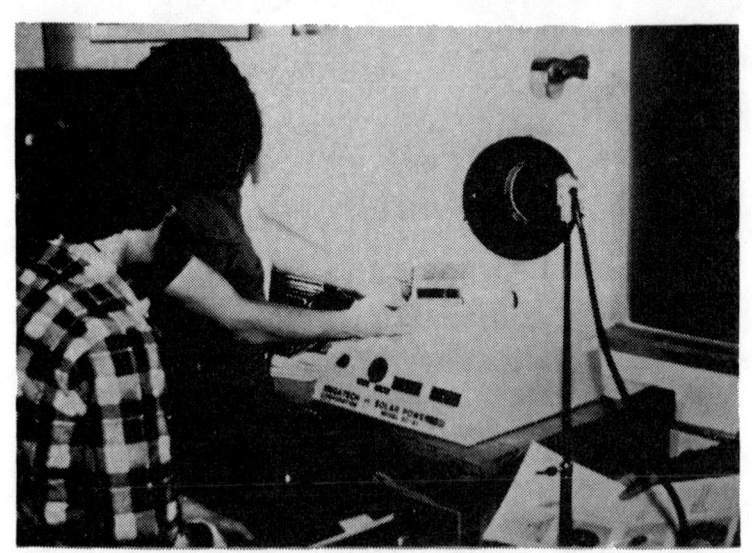

Fig. 2-12. Using artificial light to produce voltage with photocells (courtesy Megatech Corporation).

the solar cells require very little maintenance is one of the main reasons that they were adapted for large-scale use in the space program. Though this process is somewhat difficult to understand, the solar cell is really a very simple device and could have a tremendous impact in the years to come. Solar cells may become as common tomorrow as transistors are today.

Solar Electric Generators

3

We have discussed how the cells are made and the theory behind their operation. It is also important to understand how the cell actually performs and how cells can be used in powering electrical devices. As mentioned, one cell by itself won't provide much power, but much can be learned from testing and working with individual cells. From the information gained, larger solar generating systems can then be designed.

As mentioned in Chapter 2, most solar cells are silicon semiconducting devices which convert light directly into electricity. When exposed to light, each cell produces approximately the same voltage difference between its two surfaces. When a load is connected between the two contact points, the voltage difference causes a flow of current. See Fig. 3-1. This current is caused by light which is absorbed by the cell forming electron-hole pairs. The intensity of the light as well as the surface area of the cell has a direct bearing on the amount of energy produced. Therefore, cells with smaller and larger surface areas are produced to fill the various energy requirements.

THE CELL

Cells come in many shapes and sizes. Some cells are round; some are half round, quarter round, or eighth round. Some cells are so small they can be used to charge batteries in watches or calculators when exposed to light from the sun (see Figs. 3-2 and 3-3). Even artificial light can produce enough electricity to recharge the

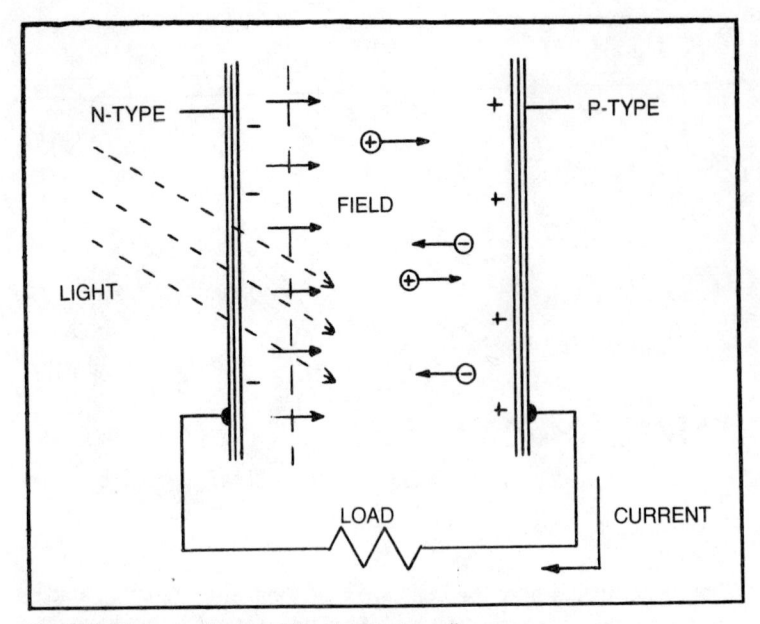

Fig. 3-1. Diagram of current flow in a solar cell.

batteries. These small cells are called *microgenerators*. Hexagon cells have been made by some manufacturers to reduce wasted space between the cells when they are placed in modules. Space cells were rectangular for the same reason. But in order to make the cells any shape other than the shape of the round ingot from which they are cut, the circular disc must be trimmed, resulting in more waste of materials. Consequently, most of the cells are round. To make cells into halves or quarters, the round cells are scribed and broken. The advantage of the smaller sections is when less current is needed, they can be arranged in smaller spaces with fewer gaps. Each cell, no matter how large or small, produces the same voltage. But a small cell will not supply the *amperage* of a large cell. If a higher current is desired, larger cells should be used. The voltage is independent of the area. Most cells deliver about 0.45 volts of power, so in order to reach higher voltages, the cells must be connected in series. The voltage is determined by the potential difference from one side of the cell to the other, and that is why all cells have the same voltage characteristics. The current, however, is directly proportional to the amount of surface area of the cell exposed to light. This is the reason a small cell will produce less current than a larger cell.

As we examine a typical volt/ampere curve, we can readily see some of the electrical performance characteristics of a solar cell. As

Fig. 3-2. Various sizes and shapes of cells (courtesy Sensor Technology).

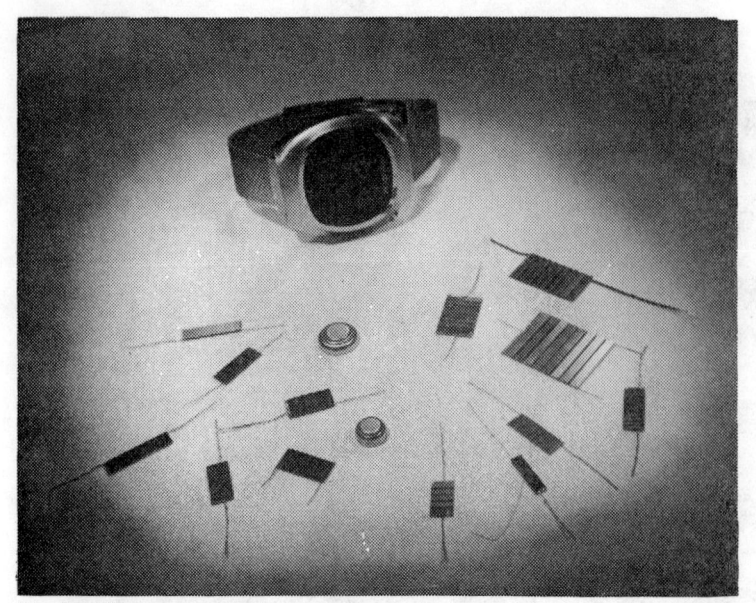

Fig. 3-3. Microgenerators (courtesy Solarex).

the intensity of light changes, the current also changes, while the voltage remains the same. The operating points shown on the curve in Fig. 3-4 will deliver maximum cell performance.

A solar cell is nothing more than a large-area diode. The individual characteristics of the cell are directly related to the voltage-current characteristics of a diode. By adding the photo current to the normal curve, the current-voltage behavior of the solar cell is obtained. The power available from the cell can be seen in Fig. 3-5. Notice the similarity of the voltage-current polarity of a cell with that of a battery.

In the electrical world today, the most common term is watts of power. To find the number of watts a solar cell produces, simply multiply the amount of current by the total voltage of each cell. When purchasing a number of solar cells, you pay a price per peak watt. This means that at noon on a bright day, the cell will reach its peak performance. On cloudy days, the watts delivered by the cells will be far less than peak. So you pay for maximum watts expected from cells under ideal conditions. When buying solar cells, consideration should be given to the fact that solar cells last almost indefinitely if properly installed and cared for. Unlike the common lead-acid battery which will last only a certain length of time, solar cells will last virtually indefinitely in their electrical production qualities. There is nothing destroyed or lost as the solar cell produces electricity.

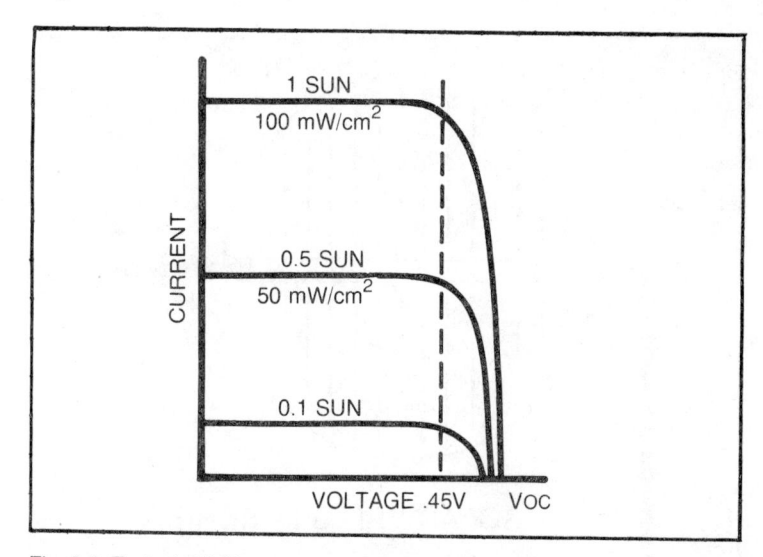

Fig. 3-4. Typical I-V Characteristic of Solarex solar cell.

After examining each cell, we next want to see what connecting several cells together will do. We know that each solar cell is capable of producing only a given amount of power. The output of each cell is very small by itself. So, if more power is needed, more cells will be required. Putting many small cells together is like getting several people to push a car that moves easily along, although one person couldn't even budge it. In the horse and buggy days, one horse could

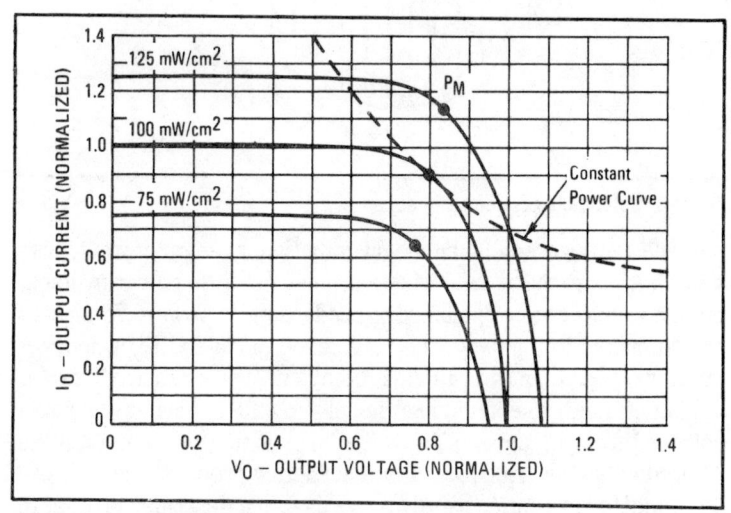

Fig. 3-5. Typical performance curve (courtesy Motorola).

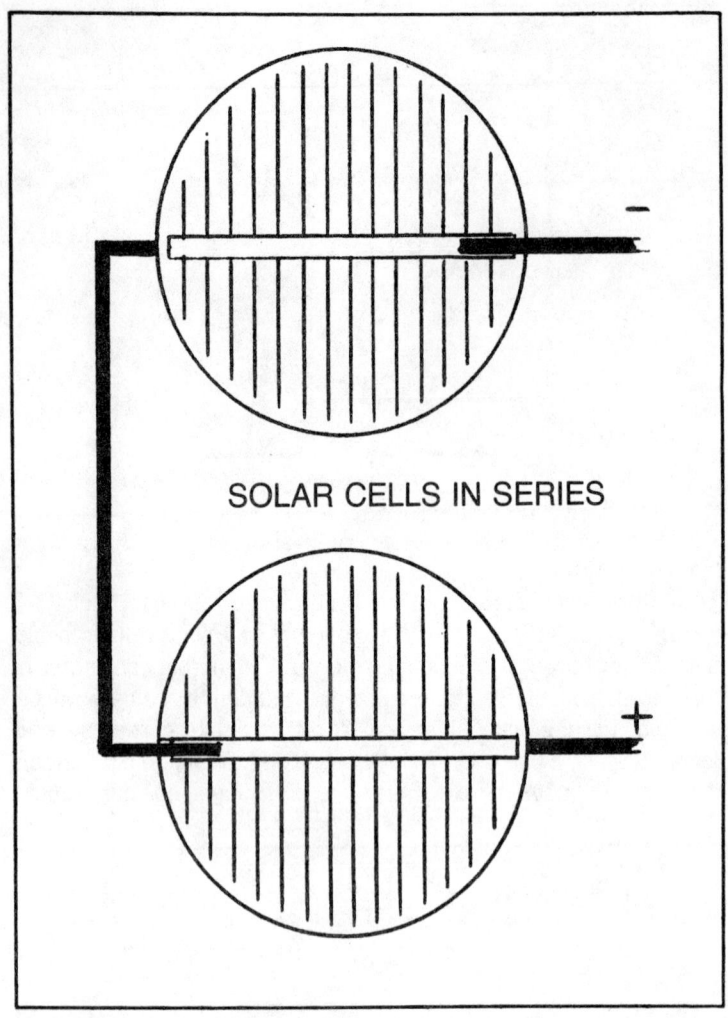

Fig. 3-6. Solar cells connected in series.

pull a certain size wagon, but if you got a bigger wagon, it would take more horses to pull it. Likewise, each cell has little power by itself, but many cells can be combined to produce more power. When cells are combined in this way, they are called modules. So, whenever greater power than can be delivered by one cell is required, a solar module is used. In connecting the cells together, each cell is hand-soldered from its upper side to the base of the next cell until the desired voltage is produced. In the soldering process, the cells can be connected in a series pattern as we have just described in order to increase *voltage* output. This is shown in Fig. 3-6. Or they may be connected in parallel where the leads from the top and bottom of one cell are connected to the top and bottom leads, respectively, of the next cell to produce a higher *current*. A parallel connection of cells is shown in Fig. 3-7. By connecting the cells in a series or parallel fashion, or a combination of both, any amount of voltage or current can be obtained. The total voltage is equal to the sum of the voltages

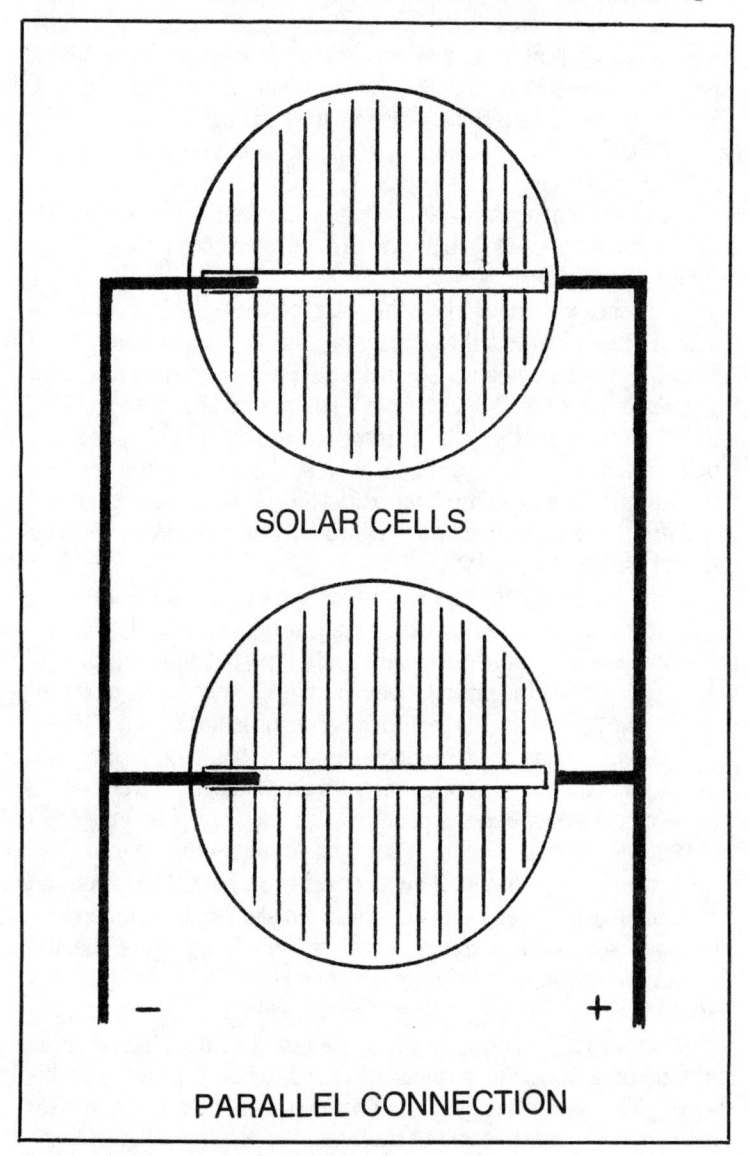

Fig. 3-7. Solar cells connected in parallel.

from all cells when connected in series with the current remaining the same as for one cell. In a parallel connection, the total current would be equal to the sum the currents from all cells current, with the voltage remaining the same.

For example, if you connected 30 to 40 2-inch cells in a series arrangement, you could produce enough electricity to trickle charge a 12-volt battery. The battery would require recharging once a day. Where more current is needed, several of these modules can be put together in parallel and still give 12 volts, but *more power*. Contact points or terminals are conveniently placed on each module so they can be joined to other modules to get the desired power needed (see Fig. 3-8). Solar electricial power systems are made from basic building blocks having a certain voltage and output. The solar modules, or building blocks, may in turn be connected together to produce even greater power.

All of the cells which make the solar module must be connected together and mounted in such a way as to protect them from damage. After the module is constructed and the cells connected together, it must be tested to see if it has electrical continuity. This must be done before the cells are mounted into the module. After the module is completed, repairing a poor solder joint would be very difficult. Each joint must be checked and the entire module examined to ensure that each contact point is properly placed before the cells are encapsulated.

The capsule completely surrounds the cell, protecting it from the environment and also providing a means of mounting. To do this, the modules are constructed using some type of frame which provides rigidity as well as means for mounting. Contact points mounted on the frame are connected to the leads coming from the cells. These contact points usually protrude underneath the module and are generally covered by the mounting frame so all the wires can be interconnected between the panels where they are protected from the weather, as well as being out of sight. A piece of tempered glass is then fitted to the frame. The solar cells are placed between the frame and glass. To prevent the cells from slipping around, and for additional weather protection, silicon rubber is normally injected between the cell, glass and frame to completely fill the voids. Small pieces of silicon are used for spacers to allow the silicon to completely surround the cells. Also, loss of efficiency through oxidization of the metal grid on the surface of the cell can be prevented in this manner. The glass glazing in front provides protection and allows for maximum light transmission to the cells. As shown in Fig. 3-9, the silicon rubber completely envelopes the cells, providing a protection

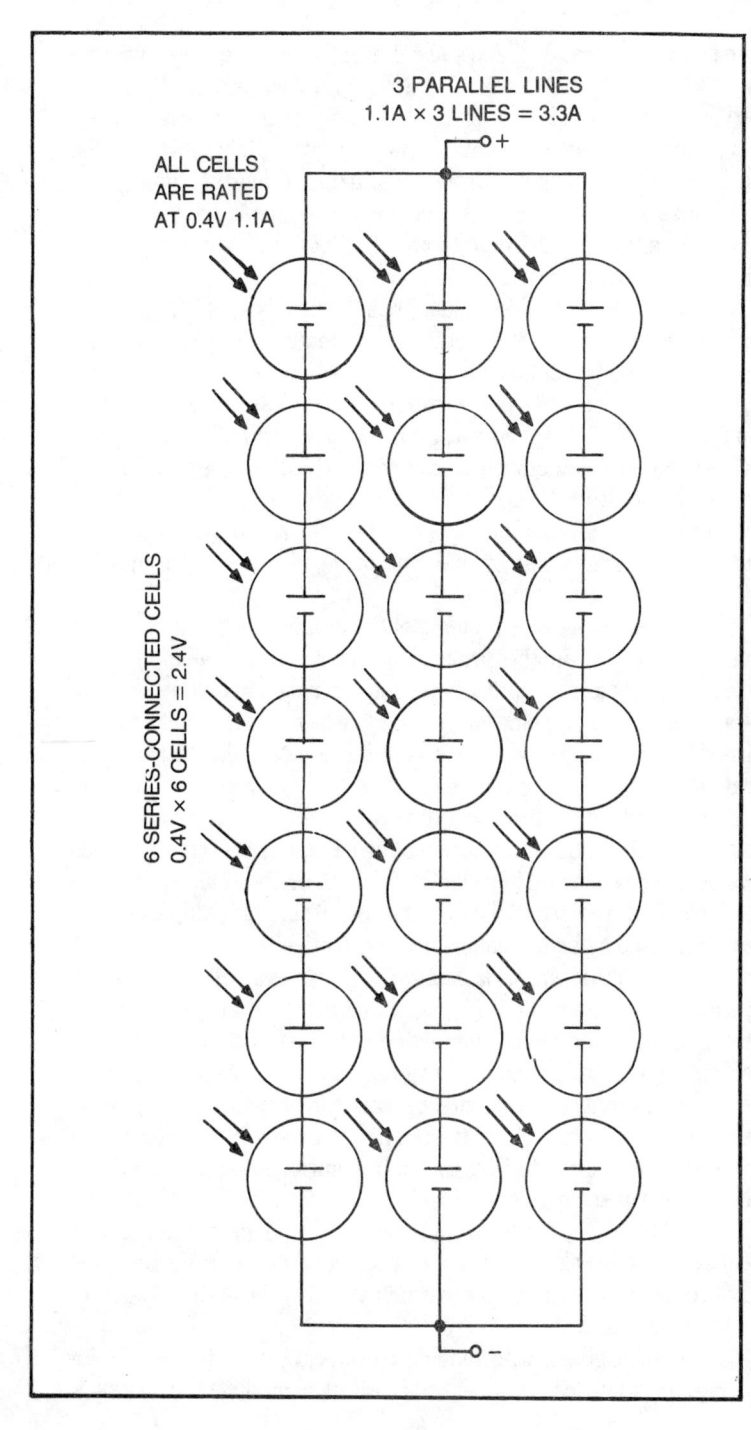

Fig. 3-8. Series-parallel connection of solar cells to provide any voltage and current rating. These cells are rated at 0.4 volts at 1.1 amperes.

from shock or vibrations, and seals the cell to exclude moisture from the module. In some cases, plastic or fiberglass has been used as glazing. However, glass is preferred in most cases because it has proven to be moisture, dust and impact-resistant. It is a material that has had a long history of survival in different environments.

These modules must be designed to withstand any environmental condition to which they may be subjected. Where danger of hail stones or rocks exists, high-impact glass is used. When exposed to sea water or corrosive substances, stainless steel frames are needed. Due to the way the modules are constructed, a very high life expectancy can be obtained.

After the cell has gone through the process of encapsulation into the module, it is again tested for reliability and efficiency. Also, any blemishes or defects which have developed in the encapsulation process must either be corrected or the module will be unacceptable. Sometimes tiny air bubbles will be left in the silicon rubber. These can be driven out by injecting more silicon into the bubbles with a hypodermic needle.

One of the inherent problems of the modules is that a cell might have been broken in the process of constructing the module. A broken cell acts like an open circuit and can result in no power being obtained from any of the cells. The module behaves much like the old Christmas tree lights; when one bulb burned out, the whole string went out. A cell having only two connections, one front and one back, could be separated by a break between connections. This would affect the entire module. This can be quite a loss after the cells have been encapsulated because it is next to impossible to get to the cell to repair or replace it. To prevent this problem, multiple contact points are used by some manufacturers. Notice the six-point configuration used in the Motorola cell in Fig. 3-10. If a cell gets broken, the electricity will be shunted around the break through one side of the cell or the other where connection remains intact eliminating the problem of a short circuit. So, when a cell is fractured or broken, there are still areas where contact points can provide flow of power. This type of cell is almost unaffected by the break, because electricity can still be harvested by the entire surface of the cell. If the contact point were only on one side of the cell, the side which was disconnected by the break would cease to produce any electricity because of the break in continuity. The redundant interconnections of photovoltaic cells increase reliability and reduce the chance of electrical losses.

Another problem with modules producing electricity is that of shading. In some cases when one cell of a module is shaded, it

Fig. 3-9. Module enclosed in plastic (courtesy Silicon Sensor).

becomes to a degree an insulator instead of a conductor, greatly reducing the amount of electricity flowing through that cell. If the cells in the module are connected in series, all the electricity from one cell must flow through the next, and so on. Therefore, a shaded

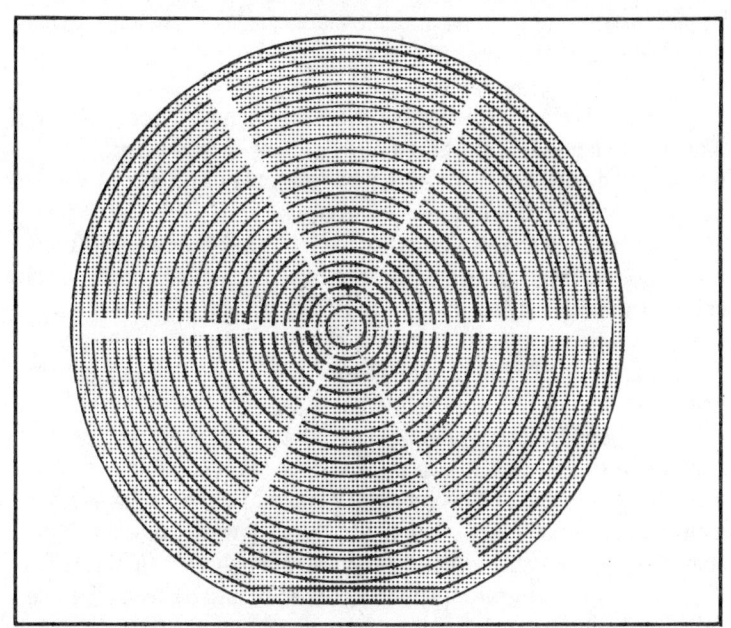

Fig. 3-10. A six-point configuration on a solar cell (courtesy Motorola).

cell acts as a resistor, letting through only a small amount of power. It is much like pinching a hose as water runs through it and thereby decreasing the flow. Multiple interconnections help to lessen this problem; also, using some parallel connections as well as series tends to reduce this problem. But ultimately, the module needs to be mounted in areas free from obstructions which could cast shadows.

LARGE SOLAR ARRAYS

Several modules are connected together to form a large unit where more power is needed than one module can supply. Such an array is shown in Fig. 3-11. These interconnections of modules into larger configurations are called arrays. The assembly of solar modules into arrays is used to develop power for a particular application. Because of the flexibility of being able to wire the modules together into arrays, almost any conceivable amount of current or voltage can be produced. The module, then, is the building block of the photovoltaic system. Modules, like bricks, can be placed together, forming larger structures of the desired sizes. The need for uniform panel sizes and standardized mounting procedures in making an array is apparent. Solar arrays are shown in Fig. 3-12.

Wiring of cells in the modules, or making modules into arrays can be done either in parallel or series, depending on the voltage or amperage required. Most modules are made to match existing batteries and power systems. Modules of 1½-, 6-, 12-, and 24-volt configurations are available.

When making large arrays, special hardware for mounting modules is needed. Many times arrays are placed on existing structures such as buildings. But many large arrays in remote areas are made on free-standing frames. By proper planning, these structures need not hamper the use of land area for other purposes. If these arrays are mounted on poles, for example, land beneath could still be used for grazing cattle. One of the advantages of solar cells is that they don't require as much land area as do thermal power plants. Thermal power plants must be fenced off for protection from animals.

Another distinct advantage of photovoltaics is the item of maintenance. Most solar panels require very little, if any, maintenance under normal environmental conditions. The solar array is almost maintenance-free. It is natural that now and then the light-collecting surface should be cleaned. The amount of light entering the array is directly proportional to the amount of electricity received. Thin layers of dirt, dust or ice will not decrease the efficiency very much. Rain usually keeps the panels clean naturally. Most

Fig. 3-11. Several modules connected to form an array (courtesy Solarex).

systems are designed with self-cleaning surfaces. However, during periods of no rain or in areas of little rainfall, panels may need to be washed down periodically. Heavy layers of dirt, dust, ice or snow may reduce the efficiency of the unit considerably.

An occasional cleaning of the units will keep them operating at a high level of efficiency. When cleaning is needed, soap and water is all that is required to remove the buildup of unwanted dust and dirt, which tends to reflect or block our incoming rays of the sun. Because these systems have to operate and endure severe thermal and physical shocks, they must be made resistant to moisture and abrasion. The panels are required to be out in the weather all the time and are exposed to some of the most inclement conditions. They must be able to withstand hail storms, icing problems, extreme heat and cold, etc.

The modules should be mounted with screws or bolts, and be sealed properly with silicon rubber or some other type of waterproofing, making it possible to mount solar panels on practically any roof or framework. Some solar arrays are mounted on adjustable frameworks to enable systems to be set at proper angles on the site. Mounting hardware may include the structure frame to which the individual modules are attached. Wiring connections between modules should accessible for servicing.

STORAGE SYSTEMS

Photovoltaic systems produce electricity only when the sun is shining. Since many applications require a continuous 24-hour supply of electricity, some way of storing the energy is needed. By using storage batteries in conjunction with the solar array, this can be accomplished. Most solar electric systems require the use of a battery. Because of the need of a storage facility or batteries, most modules are designed to produce the voltage required to charge the various types of storage batteries. A solar battery charger is shown in Fig. 3-13. Applications requiring a constant voltage or current would also need a storage battery to even out the highs and lows of solar collection. Since the length of time the sun shines varies from day to day, and sometimes clouds are in the way, solar collection is not constant. The solar arrays can deliver maximum power only under optimum conditions.

A few applications can use solar-cell power directly without the aid of a battery. In these cases, the fluctuations due to the intermittent light from the sun make the photovoltaic device operate at varying efficiencies. Most applications, however, require some type of battery to provide a continuous source of power.

Fig. 3-12. An array of solar panels supplied 2000 watts for a water fountain and the world's first photovoltaic band at a national Sun Day celebration in Washington, D.C. (courtesy Solarex).

Fig. 3-13. Solar battery charger (courtesy Sun Tap).

The batteries used with a solar-cell charging system should be rechargeable. The battery should be at least 20 times the ampere rating of the solar array in full sun. Most photovoltaic generators use a deep-cycle rechargeable battery that is able to take large drains and high charges. A typical solar charging circuit with battery is shown in Fig. 3-14. When solar cells are used to collect the sun's energy and it is stored in a battery, a *blocking diode* is needed to prevent the battery from losing its charge back through the solar cells at night. Otherwise, the cells would drain off the power they had supplied during the day whenever the sun wasn't shining. Soon a state of equilibrium would result. To prevent this loss of battery power, the diode is connected between the solar array and the battery. Placing the diodes in this way allows the current to flow in only one direction. This diode can be referred to as a blocking or isolation diode, meaning that it prevents or stops the flow of electricity in a reverse pattern when it is in a dark condition.

BATTERY VOLTAGE REGULATORS

Another item needed in a total system is a *voltage regulator*. Whenever batteries are used, a voltage regulator is needed to control voltage delivered to the battery, as shown in the solar battery charger schematic in Fig. 3-15. This device prevents overcharging the battery during full periods of sunshine. This would

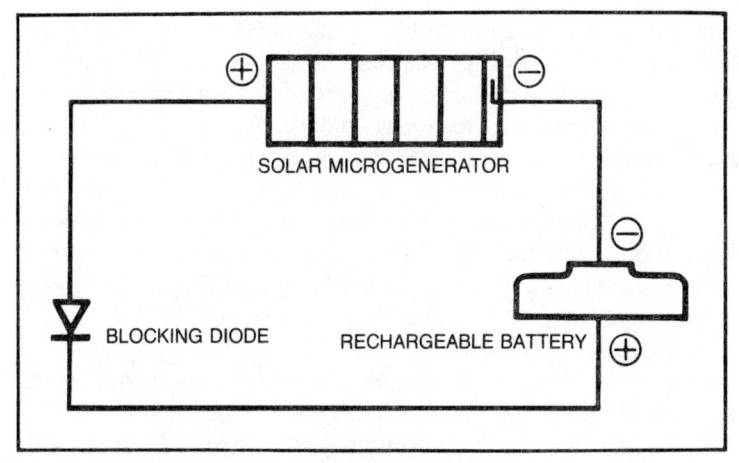

Fig. 3-14. Solar power circuit (courtesy Solarex).

normally occur when the battery reaches its full charge and the array continues to supply energy. In such a case, the voltage regulator would bypass the energy. Voltage regulators are primarily used in systems that are overdesigned to make up power needed during times of lower solar collection.

Once the regulating voltage is set, the regulator allows only enough current to flow into the battery as needed to maintain a full state of charge. Excess power output is then diverted to a power-dissipating device in the voltage regulator. When a battery falls below full charge, the regulators are designed to draw only enough power to operate the terminal voltage level detector (less than 1 mA). The low off-state power drain of a battery voltage regulator prevents unnecessary waste of valuable solar-generated power in

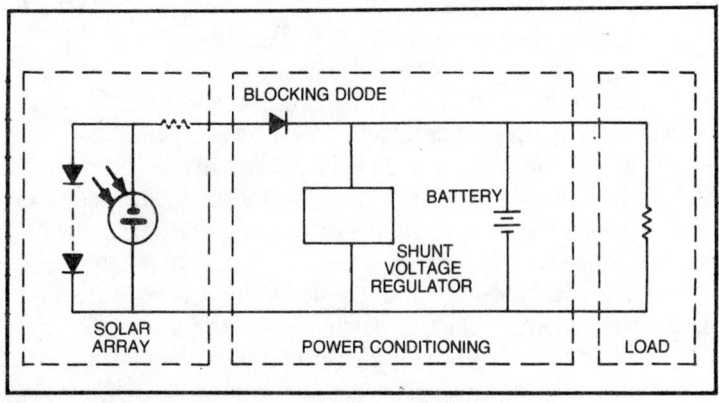

Fig. 3-15. Typical solar battery charger (courtesy Solarex).

53

regulator circuits when the battery is below full charge. The upper limit is selected to minimize the electrolysis of water in the electrolyte. Battery voltage regulators protect against equipment and battery damage caused by excessive voltage. A voltage regulator is highly recommended where the amount of energy produced by the array is in excess of the annual solar load.

The life of the storage battery would be greatly reduced without the use of a regulator. Because each application requires different amounts of voltage and power, the battery should be sized to provide ample energy storage for the periods when no sunlight is available. The generator is designed to meet the charging capacity of the battery as well as its voltage output. The battery is a buffer between the solar array and the load requirements. It supplies power to the load whether the sun is shining brightly or not. At night, or in poor weather when the sun is hidden by clouds, the load would be relying totally on the battery. The battery accepts charges from the array when the sun is shining, though the amount of charge may vary from day to day. The system should be sized so the average amount of sunshine will supply the load. The storage battery greatly aids the regulation of voltage to the load. As shown in Fig. 3-16, when installing a solar panel to a battery, the positive terminal on the panel should be wired directly to the positive terminal on the battery. The negative terminal on the panel should be connected to the negative terminal on the battery.

Probably the greatest drawback in the photovoltaic system is that storage batteries require a certain amount of maintenance and replacement. Much work is needed to improve the efficiencies and charging capacities of storage batteries. As better batteries reach the market, the total photovoltaic system will be greatly enhanced. If improved and more efficient batteries could be produced, there would be an increase in the application of photovoltaics.

A significant amount of research is being done to create good storage batteries. This is primarily because many of the utility companies are interested in batteries for storing power during off-peak hours. The cost of about $40 per kilowatt of storage capacity with a life expectancy of five to seven years is common for some of the lead-acid batteries used in the solar energy storage systems. Currently, the cost of operating these batteries is about twice what is considered desirable, but with the demand of batteries for solar application, this could change rapidly. At today's prices for solar cells, the cost of the batteries is a small portion of the total system. But as the price of the solar cells comes down, and if batteries remain at the same price, they would become the most significant portion of

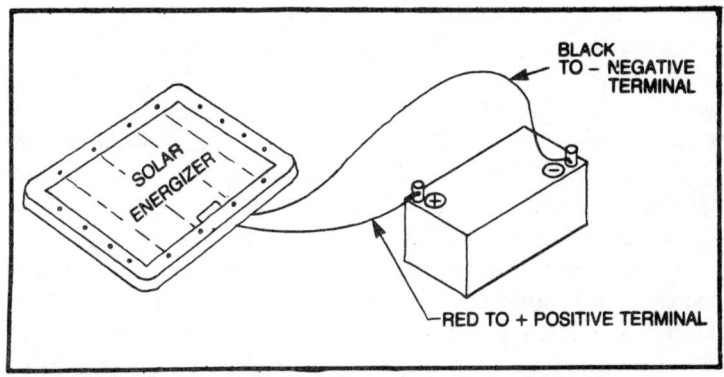

Fig. 3-16. Wiring a photovoltaic panel to a battery (courtesy Solarex)

the cost of installing a solar system. At the price of $40 per kilowatt, the storage batteries required for 36 hours of storage would cost about $200 per year, which breaks down to about 2½ cents per kilowatt-hour for the average home. An average home uses about 8720 kilowatts per year. Currently, a battery with about one-half pound lead stores about 12-watt hours of energy. Lead may eventually become a very limited substance as the demand for electrical storage becomes more widespread. Much work is now being done on other types of storage batteries. For example, experts are investigating lithium sulfide and zinc chloride. It is hoped that advanced batteries with longer lives will be developed in the mid-1980's. In some applications, the utility company power grid has been used as a storage system. Also, the possibilities of using flywheel storage and pumped storage are being looked into. If people could alter their work habits to where their use of electricity was primarily done during the daylight hours, the amount of storage needed for the solar cell generating devices would be greatly reduced. Also, the possibility of groups of people using the same storage system as a shared facility is apparent.

PUTTING IT ALL TOGETHER

Completed systems require that the solar cells be assembled into modules, which are then combined into arrays. In addition to batteries, which merely store energy as it is collected, and hold it until it can be used, diodes and voltage regulators must also be employed. All these need to be assembled and mounted in such a way as to make the photovoltaic generating system completely self-contained. The module panels should be mounted so that the solar cell is visible through the transparent glass cover. This, of course, must be exposed to direct sunlight in a nonshaded area.

The cells generate a *direct current* (DC) of electricity, which is needed in a battery system. If an alternating current is needed, as in a home, a power conditioner is needed to convert the 12-volt DC electricity into 120-volt AC. In some cases, modules have been made to produce 120-volt DC, which can be changed into AC power and fed directly into the home through the utility lines. In such power systems, the utility acts as a storage device. Using a utility as a storage system offers possibilities, if problems connected with this type of hookup can be overcome to the satisfaction of the power company. The problem now is that the solar power is produced during the daylight hours, whereas peak demands are generally in the evening hours. The utility companies have no way of storing energy, so this type of power being added to the line usually only increases the problem.

Any amount of power can be obtained by combining solar modules into larger solar arrays. If they are assembled properly, photovoltaic systems should offer a reliable, nonpolluting source of energy, designed for long-term, year-round performance for any type of prime or standby power requirements. Solar power systems then could offer a viable alternative power source to meet our ever-increasing demand for electricity. Many problems are yet to be solved to make these systems cost-effective. Much work needs to be done both in cell production and storage systems before photovoltaic power can compete with current forms of electrical power.

Efficiencies and Economics of Photovoltaics

4

In an attempt to make photovoltaics a more viable form of electrical power, the areas of cost and effectiveness must be considered. While much research is being done in developing solar cells, the biggest problem facing them is their high cost. At the present time, the cost of the photovoltaic cells prohibits their widespread use, and the fact that storage batteries of some type are required to provide electricity on a 24-hour basis for home use is an added expense. But with the reduction in production costs of the photovoltaic cells, it is envisioned that great amounts of energy can be produced with this type of device, and that many home applications may become practical in the near future.

To see how the cost of photovoltaics compares with conventional electricity generated by fossil fuels, the following example is given. A three-inch solar cell with efficiencies of about 10 percent could produce approximately 0.45 peak watts of electrical power, or about 0.082 continuous average watts. Approximately 0.718 kilowatts per year could be generated by the cell in an average location in the United States, using a cost of $15 per watt. If the solar cell produced power for 30 years, it would provide electricity at about 30 cents per kilowatt-hour, compared to the average of about four cents per kilowatt-hour by conventional sources in the United States. This shows that at present, solar power is about seven to eight times as high as that produced by other sources. Some areas in the United States, with higher-than-average power bills, may be 8 to 10 cents per kilowatt, making solar cells only three to four times as costly.

57

As costs are reduced by three to eight times the current cost, the price of solar power will then be competitive with today's prices of electricity. However, while the price of solar power is coming down, conventional prices are daily increasing, a fact which is an added advantage. As we develop new ways to make solar power more cost-effective, it is important to remember that photovoltaics, like many advanced energy devices, would have a high initial cost, but a minimum operating expense. One way to reduce the cost of photoelectricity is to improve the overall efficiency of the cells. This, then, would lessen the number of cells needed to achieve results, thus reducing the overall cost.

SOLAR-CELL EFFICIENCY

To understand the solar cells and how their efficiencies are determined, several things must be considered. By examining the materials that solar cells are made of, we find that silicon is one of the most abundant elements on the earth. But to be suitable for cell production, the silicon must be highly refined. Almost all the impurities must be removed. Achieving the degree of purity required in metallurgical silicon entails a very expensive process that greatly affects the price of the raw silicon. A reduction in the quality of silicon also has a tendency to reduce the efficiency of the cells. Other types of cells have similar problems of keeping material costs down while not sacrificing efficiency.

The principle upon which photovoltaic cells function is that the material used, such as silicon, is sensitive to only a small portion of the sun's rays. The sun's light can be directed into a spectrum of light ranging from ultraviolet to infrared. The semiconductor materials used in solar cells respond to only a certain area of the spectrum, as shown in Fig. 4-1. The efficiencies of the cell is thus determined by the area of sensitivity to sunlight.

Photons of light energy from the sun carry a minimum threshold of energy: just enough energy to release an electron within the photocell. To do this, a certain level of energy is needed. The cell cannot utilize energy found elsewhere in the spectrum. This means that only a small fraction of the sun's energy is used by the cell. Some of the light is reflected or lost through refraction. Some of the light seems to go right through the cell unused. Out of 100 percent of the sun's energy falling on the surface of the cell, only a certain portion is received and converted into electricity. The efficiencies of cells range in the 6 percent to 20 percent category. Although this seems like a small percentage of energy being converted, it must be remembered that this is a direct conversion. It is many times as

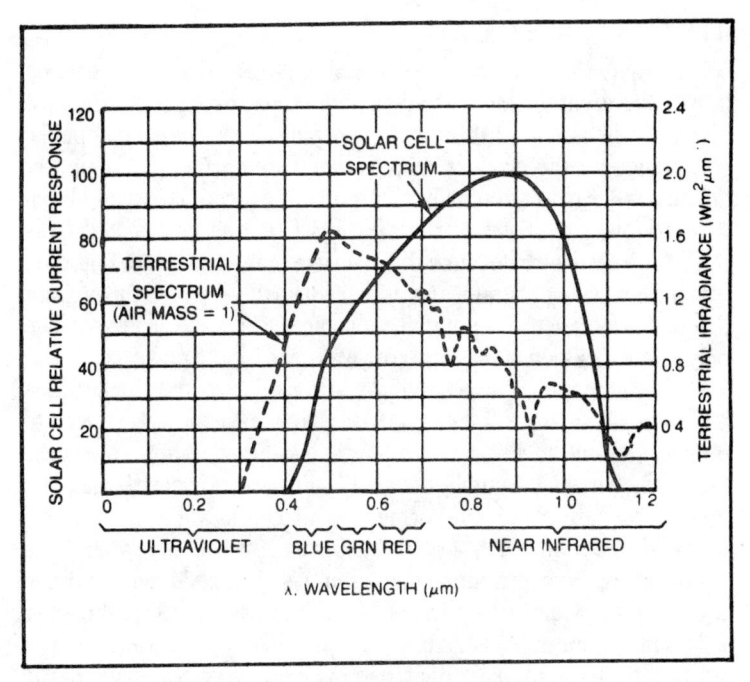

Fig. 4-1. Spectral response of solar cell (courtesy Motorola).

efficient as fossil fuels if their efficiency is measured by comparing the amount of sun received in their development to the ultimate energy delivered.

The efficiency of the solar cells is determined by dividing the amount of incoming light into the amount of power produced by the cell. Most silicon cells are achieving 10 percent to 12 percent efficiencies. However, 18 percent to 24 percent conversions have been obtained in laboratory settings. Most of the other materials used in making the cells have a lower efficiency rating than that of silicon. All the other materials currently being used in cell production also have a certain spectrum of sunlight they can use. These areas are known as band-gaps of light energy. If the makeup of the cell could be altered to receive a greater portion of light, greater efficiencies could be obtained.

Increasing the efficiencies of the solar cells could have an effect on the overall price of the cells. It requires the same basic material and labor to produce solar cells, so that those of a higher rating would cost less per watt of power produced. The problem can be solved either by improved efficiency or by cost reduction. Not only the materials, then, but also the efficiency has an effect on the cost of cells.

HIGH COST OF CELLS

Despite the tremendous potential the solar cells will have in the future, the greatest barrier to their widespread use is cost. The cost of solar cells is one of the most important factors in gaining public acceptance for the product. Like any new item, the price of photovoltaics is sure to eventually come down. Much has already been done to lower their price. Before the 1970s, the cost of cells was over $300 per watt. Because the main use of the cells up to that time was in the space program, the price remained very high throughout the '60s. This price was not objectionable in the space program since only a small fraction of the total cost was used for the power devices.

Even though the use of photovoltaics in terrestrial applications has come about since the early 1970's, they have already proven to be reliable and successful in many applications. When solar cells began to be used in a number of applications here on earth, the price began dropping. The price came down to $100 per watt in 1970, dropped to $30 in 1973, and by 1975 was down to as low as $17 a watt. One of the contributing factors in this price reduction has been lower-cost modules using larger diameter cells. In 1976, the solar cells suitable for use on earth were priced between $15 and $20 per watt, showing a dramatic decrease over prices of cells used in the space industry. The space cell had to be of higher quality because there was no way to repair a faulty panel in space. By 1977, prices had been reduced to about $15 per watt. At this price, the cost of generating photovoltaic power is still 10 to 15 times the cost of generating power from coal. Even still, this reduction has brought about a number of new products using solar cells to supply their power. Of the $10 to $13 per watt, one half represents the cost of the solar cell itself, and the other half represents the cost of putting the cell into a panel.

As the industry grows, silicon producers could make a product especially formulated for the solar cells at a much lower price, greatly reducing the cost of the solar cells. Because of the small quantities of silicon purchased for cell production compared to that of other semiconductor devices, the same quality of silicon is used for both. But as more silicon is used for cells, a material less highly refined could possibly be used. The semiconductor grade silicon is 99.999 percent pure and requires a costly process to make it that pure. Because of this, it costs about $10 to $30 a pound. If solar cells could use a silicon which is about 98 percent pure, the cost would be 10 to twenty cents a pound. This may produce a less efficient cell, but the trade-off in price may be justified.

There are other reasons why photovoltaics are so high in addition to the fact that raw silicon is so expensive. Waste is involved in cell production. A completed solar cell receiving the final test either functions or not. Any defective grid or crack in the cell would cause it not to operate properly. Many cells are rejected. Perhaps the cell works fine, but is inferior in appearance; this too may be rejected. The waste in these examples, plus those of manufacturing, has a tendency to make the cells cost more.

Another example is the method by which the cells are made. At the present time, most of the manufacturing procedures are accomplished through hand processes. The cells are cut and etched by hand. Wires soldered by hand, connections are tested by hand, and so forth. The labor accounts for almost 70 percent of the total cost. Finished cells must be mounted into a module by a hand-soldering process. If automobiles were produced in like manner, we would pay 20 times more for them than we do today.

Another reason cells are so high-priced now is that they are not enough in demand. The volume of cell production is still so small that it has not justified the cost of automating production facilities. The total production of solar cells in use to date is about 150 to 200 kilowatts, or enough to power approximately 40 average United States homes. Because of low production and limited market, prices remain high. In the future, as new technology and automation methods are developed in the manufacturing process, the price of solar cells is expected to decrease. If production could be scaled up sufficiently to permit automation, manufacturers, researchers and government officials all agree that the high cost of solar cells would drop accordingly. Because the solar industry is so new, studies into other closely-related fields justify hopes of future decreases. A Boston consulting group conducted a study in 1970 and found a relationship between volume increases and price reductions. As the volume doubled, the cost dropped as much as 20 percent to 30 percent. This was brought about partly by automation and partly by improving production techniques. With improvements in the manufacturing and methods of quality control in the silicon-cell process, the price could be reduced drastically.

Solar cells would lend themselves easily to mass-production techniques because they are an electronic product. One example of this process is the closely akin transistor industry. When the Department of Defense needed a light weight electronic replacement for vacuum tubes, transistors were developed, initially costing about $20 each. But, as the demand increased, prices went down to 20 and 30 cents per transistor. The same type of situation is expected to occur with solar cells. Photovoltaic cells would actually be much easier and simpler to produce than integrated circuits. An integrated circuit is a solid state device which takes the place of many transistors and other components by combining them. Production of cells requires only 100 steps, whereas the fabrication of integrated circuits requires over 500 individual processing steps. As new methods for the use of solar cells and new products are sought for and found, prices will continue to decrease (see Fig. 4-2).

Assuming current technology continues with no major breakthroughs, and volume only increases sufficiently to justify automation, the price of solar cells could drop from $20 to $2 a peak watt by 1980. There are strong reasons to believe that within the next five years, the cost of photovoltaic devices will come down to the level of $1 to $2 per installed watt. It is hoped that this will take place in the early 1980s. At this price, photovoltaics will start to be economically feasible in some market applications.

Let's consider an example of what it would cost at this point if a home with a 700 square foot area of roof were covered with cells. Using an overall efficiency of five percent, 3 kilowatts of power could be generated in about five hours of average sunshine. This would mean about 17 kilowatt-hours per day. This is quite a substantial amount of electricity from one portion of roof area. At a price of $2 per kilowatt, our system would cost over $7000. Even at this price, a system of this type would make a lot of sense. This, of course, will change as the demand for cells increases. But imagine what could be done if prices were cut in half or even down to 50 cents per watt. It is thought that by 1990, prices may be down to 30 cents to 50 cents per watt. When prices reach 30 cents per watt, solar power will be comparable to the amount that many people are paying for electrical power from conventional sources at today's prices. Who knows what electrical power might be costing by then?

As prices of conventional power increase and prices of solar cells decrease, the time when solar cells will be competitive will be shortened. Somewhere along the line, solar cells will become a very attractive alternative to conventional forms of electrical power. While the initial cost remains high, the operating cost for photovoltaic systems is relatively low. This system requires no fuel or maintenance, so the initial cost is basically the only cost involved.

FUTURE GOALS

The goal of the photocell industry has been to get the price down where they can be competitive with today's other sources of power. This can be accomplished in several ways:

Fig. 4-2. Small items, such as this solar-powered calculator, help create a demand for photovoltaics (courtesy Solarex).

- ☐ by reducing the cost of raw materials through the use of a less pure raw material or through quantity purchases to obtain better price cuts;
- ☐ by improving processes by which cells are produced, such as automating many of the hand operations;
- ☐ by improving the quality of the cells and increasing their yield, as much profit is lost to the waste of broken or unusable cells;
- ☐ by improving cell efficiency to utilize a greater amount of the sun's energy;
- ☐ by finding some other cheaper material or better method of producing the cells.

In attempts to reduce the costs of cells, many manufacturers and laboratories are conducting a variety of experiments. One is an attempt to superimpose several cells over each other to improve overall efficiency. These cells each separately collect a different area of the sun's spectrum. As the light travels through these thin layers, a greater amount of the sun's energy is absorbed. In this type of cell, efficiencies of up to 50 percent have been obtained in laboratory settings. The cost of producing this kind of cell is greatly increased, so whether or not its improved efficiency offsets the increase in price will remain to be seen.

Another area being developed is producing silicon cells in a continuous manner, forming a ribbon. This procedure eliminates time involved in growing crystals and avoids the waste involved in cutting cells into wafers. This process could lead to mass-production techniques needed to reduce the production costs.

Several new materials are being tested, many of which are getting higher efficiencies than silicon. Thin film cells, made by depositing a semiconductor material on plastic film, could be the answer because of their ability to be mass-produced and because of their simpler process. These new materials and processes will be discussed in Chapter 7 in more detail. All these developments will help reduce the cost of the solar cells in time, but the paradox is this: As long as the cost of cells is high, the demand will remain low. The annual production will be limited to a few hundred kilowatts, and small companies involved will not have money to automate. So the very things needed to bring costs down are hampered because of low demand. However, the demand may change as conventional power escalates.

Because of the current high cost of solar cells, and small production capabilities of the present industry, the limited demand will prevent photoelectric power from playing a significant role before the end of this century. Unless something can be done to accelerate the growth of the industry through improved developments and greater demand, the market for photovoltaics will not be sufficiently large to induce investments by private industries. But market indications show that as the costs of solar cells are reduced, the existing industry of one or two dozen small manufacturers can be expanded dramatically. The industry could expand both in volume and types of products employing solar power. It is expected that photovoltaic arrays can find wide commercial use if prices can be reduced to $100 to $300 per peak kilowatt. Realizing this problem, the federal government has become involved in trying to create a demand for the cells to justify manufacturers' expense in automating.

"The primary goal of the Energy Research and Development Administration (ERDA) Solar Energy Program is to develop and demonstrate as early as practical those terrestrial solar energy applications that are commercially attractive and acceptable." The current photovoltaic situation is accurately described in this statement: "As a result, the federal role has been accelerated in the utilization of photovoltaic systems, and a basic plan has been adopted to cause photovoltaic systems to become so attractive that they can provide electricity for key markets by 1986, and a significant percentage of the United States demand by 1990. The photovoltaic program plan is a multi-year plan structured to rapidly reduce the costs of systems and to promote rapid expansion and production and the use of photovoltaic power systems. Photovoltaic systems are modular in design and an expanded production capacity for low-cost systems will open a wide spectrum of markets that range from small systems to central power stations including agricultural, residential, service, commercial, institutional, industrial and utility applications." The Department of Energy has funded several large installations to stimulate the industry.

At today's prices, cells are limited to only those who cannot economically link up to the conventional utility grid because of remote locations. If array prices could be reduced by a factor of 20 to 30 times the present costs, photovoltaics would become competitive in today's market. As demand increases and price decreases through automation, and as a result of the use of thin films and new improvements, it is hoped that the industry will become attractive to private investors.

Applications of Photovoltaics

5

Whenever people live and work in remote areas, whether on deserts, in canyons, atop mountains, or along coastlines, they need some form of energy. And today it seems that one of the most useful and acceptable forms of energy is electricity. But many times, this convenient form of power is not available to people from the utility networks now linking most populated areas of the United States.

In remote areas where power lines haven't yet reached, the cost of running new lines is often prohibitive. In such cases, other types of power generation are needed if electrical power is to be used. Diesel generators, windmills and waterwheels have all been used, each with inherent advantages and disadvantages. Because most of these are highly mechanical, maintenance becomes the major problem.

There is, however, another source now available to them: photovoltaics. This solar electric power is often becoming preferred when line power is not available or not economically feasible. Not until photovoltaic cells were employed to generate power in these remote generators. In an article in the *Mother Earth News*, Copthorne Macdonald wrote, "Wood may be burned for heat and candles used for light, but in this day of electronic devices, there is no substitute for electricity." The solar cells have made it possible to have a reliable source of electrical power in almost any remote place in the world. They produce electricity silently and efficiently whenever the sun is shining. They require practically no maintenance and will operate almost indefinitely.

As yet, solar sells have not solved all the remote electrical problems. They still have a high initial investment for installation. However, it has been found that a three-kilowatt-hour per day photovoltaic system often operates more economically than motor generators. Wherever a modest power demand exists and other power is not available, solar electrical power is already cost-effective or will be in the near future. Installing solar power stations in some remote areas can now realize potential savings, or additional power can be supplied where existing power sources are limited. Most of the photovoltaic systems constructed to date are being employed in isolated situations where batteries are impractical or difficult to recharge, or where other sources of power are to expensive to obtain.

The National Aeronautics and Space Administration (NASA) has used solar power successfully on many of their space flights. Photovoltaic power has been used on space crafts to provide power for electronic equipment since 1958. *Mariner 5*, powered by a 325-watt solar array, took some of the first photographs of Mars and returned them to earth. These systems vary greatly from those needed for terrestrial applications, but they have been an excellent source of providing the technology and experience needed to develop the new terrestrial cells in use today. Along with the many applications in space, many successful examples have taken place on earth as well in recent years.

Until 1970, most of the cells available for earth use were rejected space cells. But by 1975, over 100,000 watts of electrical power was being provided by modules on earth. This was a landmark year because for the first time more solar cells were being made for terrestrial uses than for use in the space program.

Economics points to the fact that the most efficient use of silicon colar cells would be that of decentralized power applications (see Fig. 5-1). Because power could be produced right at the source of the demand, it could eliminate the need for transmission lines used with conventional power sources today. It has been calculated that the average roof of a home in the United States could produce enough power to supply most of the home needs with sufficient power remaining to keep an electrical car charged up. About 500-550 square feet of roof area would be required to generate approximately 1 peak kilowatt of power in most areas of the United States. In some localities where *insolation received solar radiation* is greater, such as the sunny Southwest, this area could be reduced by 25 percent. However, in many urban areas with large apartment complexes, there is simply not enough roof space for solar arrays to provide

Fig. 5-1. Solar-powered sanitary facility at Custer National Forest (courtesy Solarex).

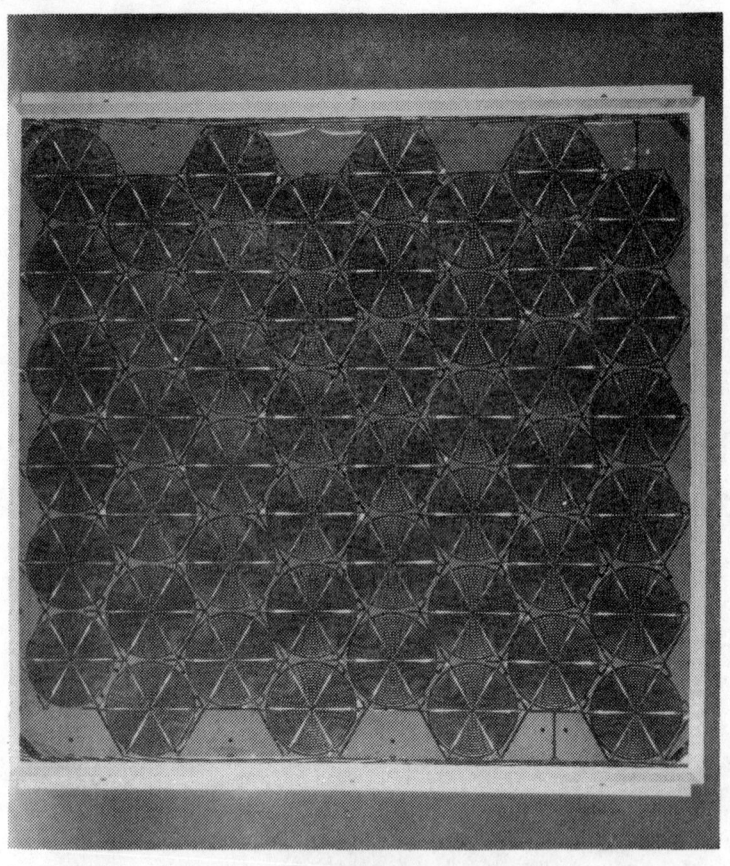

Fig. 5-2. Photovoltaic module (courtesy Motorola).

ample power to each dwelling. In these cases, centralized power stations would probably be a necessity.

Solar cells are also being used in small electronic applications. Dozens of companies are now producing many different devices using solar cells. Each application requires different sizes of cells or modules. Because of the variety of applications, modules are made in a wide assortment of sizes and shapes. They range from small chips used to run watches, radios and calculators, to many large cells grouped together to produce electricity in huge solar power plants (see Fig. 5-2).

It appears that as prices decrease, more uses of solar cells will be found. We will attempt to discuss some of the applications currently being used or explored. In some areas where it is too costly to lay electrical cables, solar arrays are providing power. In these locations, solar modules are now being used as a primary source of electrical power when power requirements are relatively modest. Whenever primary batteries are used and thrown away after they are discharged, solar cells are already competitive. The remote market applications for photovoltaic cells are increasing daily. The use of solar cells now and in the future is and will continue taking on new dimensions. Today's commercial market will see an increased use of solar cells on such small items as toys, radios, calculators and remote communication systems (see Fig. 5-3 through 5-6). Different sizes of photocell panels can be tailored to fit almost any application. Solar electric power systems are becoming competitive with primary batteries, fossil-fuel-fired thermoelectric generators, and small-motor generating systems. But the photovoltaic market is limited to relatively small power applications, usually less than one kilowatt. In remote locations such as deserts, mountains and shore lines, these power units become viable installations (see Fig. 5-7). Because of the unique characteristics of the solar cells, they are now being used in a greater variety of applications. In the next two sections of this chapter, photovoltaic applications will be discussed. For purposes of clarity, these have been divided into two groups:

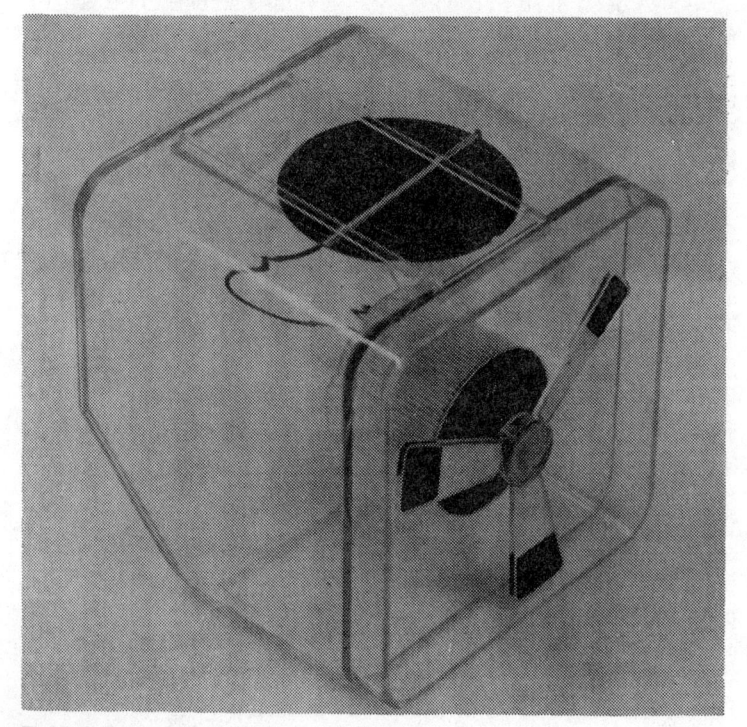

Fig. 5-3. Solar cube toy (courtesy Solarex).

first, smaller applications; and second, larger installations. As some of the uses are pointed out, no doubt many of the readers will see other applications where the cells could be employed. There is a whole new frontier in this field where inventors and market people can find avenues of possible financial reward.

SMALL SCALE APPLICATIONS

Solar cells have been used in a variety of applications. They have helped to catch insects for pest control research at Texas A & M University. Solar cells have been used to monitor fighter pilot training over the Gulf of Mexico. They have also been used as navigational aids, lights, and radio communication for gas distribution platforms in the Gulf of Mexico, as shown in Fig. 5-8.

Another very interesting use of photovoltaics is the photophone. The photophone is a device used in transmitting speech. The light reflected from a mirror attached to the needle of a phonograph strikes a distant solar cell. This sound is then changed to a light source and back to sound again. These instruments have been used to keep records of the moonlight, sunshine and fog density. They have also been used to record areas of dense smoke to determine the amount of pollution coming from some factories. The application for photovoltaics seems to be very diversified, but the basic essentials of many of these applications are essentially the same.

Radio

Probably one of the greatest uses of the solar modules today is that of remote radio communication systems, (see Fig. 5-9). In relatively inaccessible locations, microwave UHF and VHF satellite central repeater stations are located on mountains, in deserts or along pipelines. The use of photovoltaic generating systems allows complete freedom of site location and eliminates the need for continually transporting fuel and mechanics to these remote sites, thus avoiding costly maintenance trips. Sun power is not only cost-effective in such applications, but in many cases, offers the only practical solution. It doesn't take much power to keep a radio operating, and this power can be supplied at reasonable costs by solar arrays. In many such cases where continual operation is needed, batteries are used to supply power at night. A photovoltaic panel supplies power to keep the battery charged during the day. A 16-watt solar array on top of the White Mountains in California is supplying power for a voice repeater station for the United States Forest Service. Also, Forest Service guards patrolling the mountains of Indio National Forest carry a 1.7-watt array backpack

Fig. 5-4. Solar-powered attic fan (courtesy Solarex).

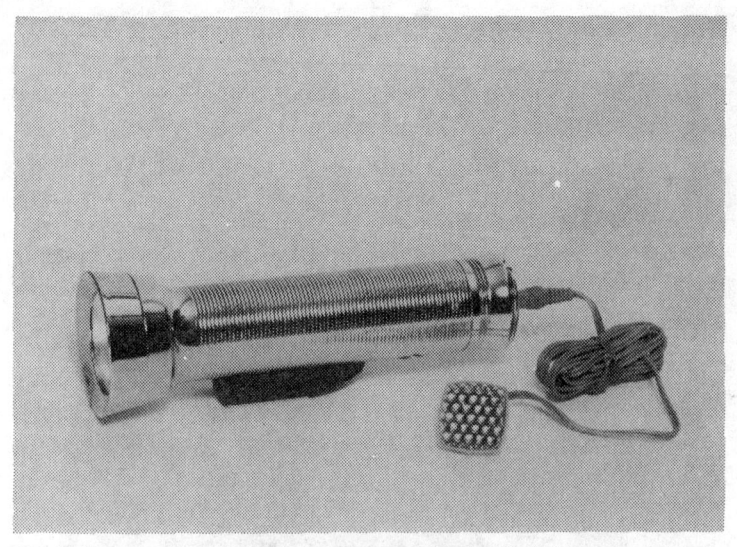

Fig. 5-5. A solar cell recharges the battery of this flashlight (courtesy Solarex).

mounted to a two-way radio. The Arizona Highway Patrol has been experimenting with solar cells for radio repeater stations and has currently given the go-ahead for statewide application. The number of installations of this type are almost too numerous to mention.

Navigational Aids

In areas of severe environmental conditions, solar arrays must be able to withstand most of the tortures nature frequently gives: hail, rain, snow, sand, wind and salt water spray (see Fig. 5-10). The reliability and ruggedness of solar arrays make them suitable for use in areas of severe weather conditions. The United States Coast Guard has a solar array that provides 50 watts of power for its navigational aids just off Long Island Sound. Entire buoy systems off the shores of Florida have been converted to solar cells by the Coast Guard. Batteries are kept charged to power lights and other marine equipment on private and commercial boats. Foghorns and bells of unmanned oil rigs, and service lights, buoys and other marine installations are now being used in areas around the world. Using cells to provide power to these types of marine applications shows tremendous savings over the primary battery-powered systems traditionally used. These solar generators are designed to operate under severe and corrosive environments, and the low maintenance required offers substantial savings over traditional battery-powered systems.

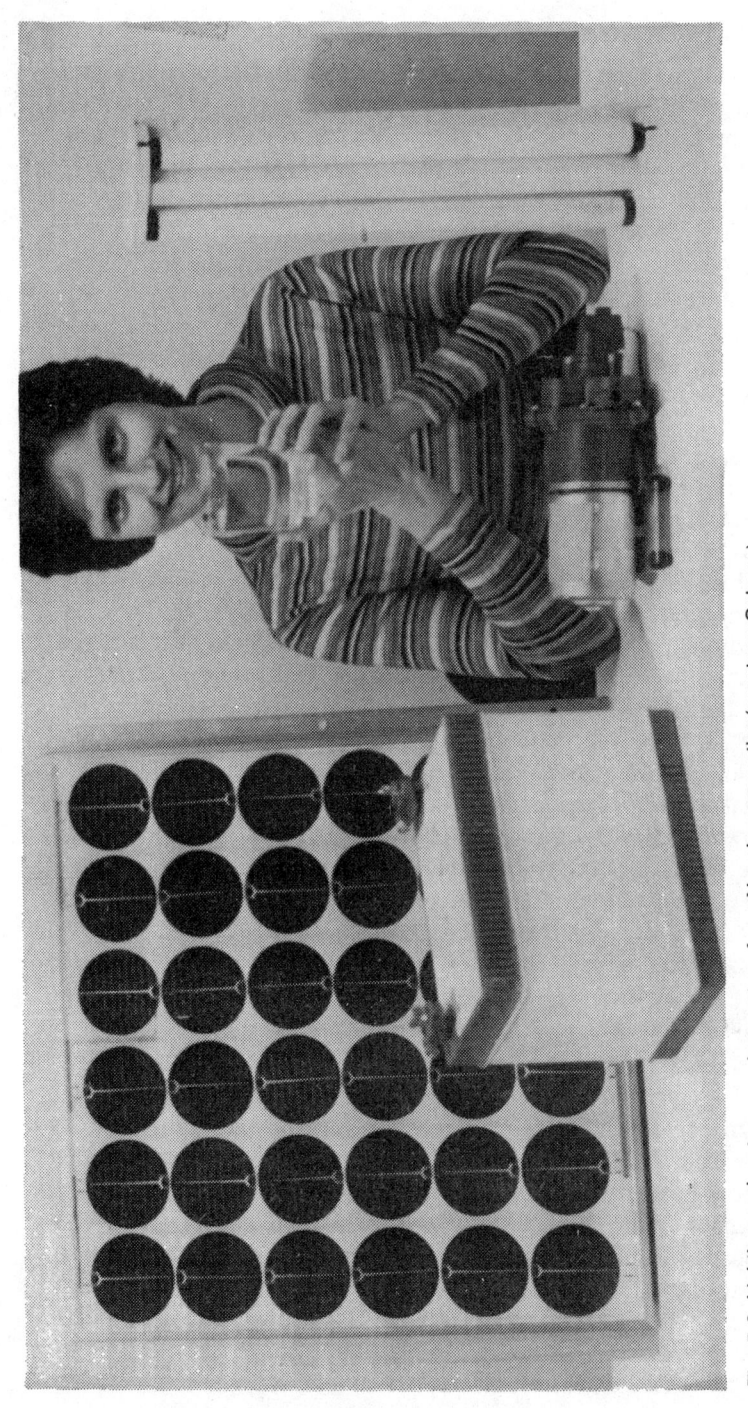

Fig. 5-6. Lighting and water pumping system for a Navajo reservation (courtesy Solarex).

Cathode Protection

Solar modules are ideally suited for cathode protection for bridges and underground pipelines, as shown in Figs. 5-11 and 5-12. Many fuel transmission lines and well casings are catalytically protected by solar power systems which provide power to neutralize the static buildup of electricity. Without this protection, static electricity causes a corrosive action which can destroy the pipelines in a period of less than five years. With this protection, the life of pipes has been extended to more than 25 years!

One of the best ways to keep bridges from becoming slippery on cold, wet wintery days is to put salt or other deicing chemicals on the roadway. However, these chemicals cause serious problems. As they are dissolved by water, they seep into the cement on the bridges, causing small static electrical currents to flow from surface of the bridge through steel or moist concrete to other locations on the bridges. This current eventually erodes the steel, causing it to swell and rust. This action cracks the concrete surface, causing large potholes to appear on the surface of the bridge, and eventually the bridge must be resurfaced or rebuilt. Highway engineers have solved this problem by using a photovoltaic unit to short-circuit the normal destructive process by attracting electricity away from the bridge through the steel reinforcement rods. Solar panels have been installed between the two George Washington Parkway bridges over Deer Run near Washington, D.C. The Federal Highway Administration has estimated savings of up to $10,000 by eliminating power lines to the sites. When sunlight hits the 90-watt solar array, electricity produced is sent to 2000-ampere-per-hour batteries where excess power is stored for nighttime use. Enough power is produced by solar cells to electrify the steel rods in the newly-constructed bridges.

Educational TV

In underdeveloped countries where people are miles from the nearest conventional power source, educational TV programs are using photovoltaic power sources to provide these services for children. In Africa, the National Broadcasting Authority of the French government is using a 33-watt panel to power a TV set in a remote school, as shown in Fig. 5-13. Over 4000 TVs with a total capacity of 200 kilowatts are being used along the Ivory Coast. In the past, the high cost of diesel or other generators made TVs in such areas impossible. By using solar cells in such applications this has become possible, and they are now very competitive for use with educational TV receiver stations. Solar generating plants are being

Fig. 5-7. Solar energy provides radio power (courtesy Solarex).

Fig. 5-8. Navigation aids and radio equipment aboard this gas distribution platform in the Gulf of Mexico are powered by 12- and 48-volt (nominal) photovoltaic arrays generating 425 peak watts (courtesy Solarex).

used because conventional remote-receiver power plants, such as battery or engine generators, are too expensive and require too much regular maintenance to be practical and cost-effective.

Railroad Crossing

Many railroad crossings in the United States and Canada are unprotected because no power is available. There are approximately 175,000 unprotected railroad crossings in the United States alone. By using photovoltaic panels, continuous reliable power could be

Fig. 5-9. Communications array (courtesy Solarex).

Fig. 5-10. Nav-aid power system used in the Gulf of Mexico (courtesy Solarex).

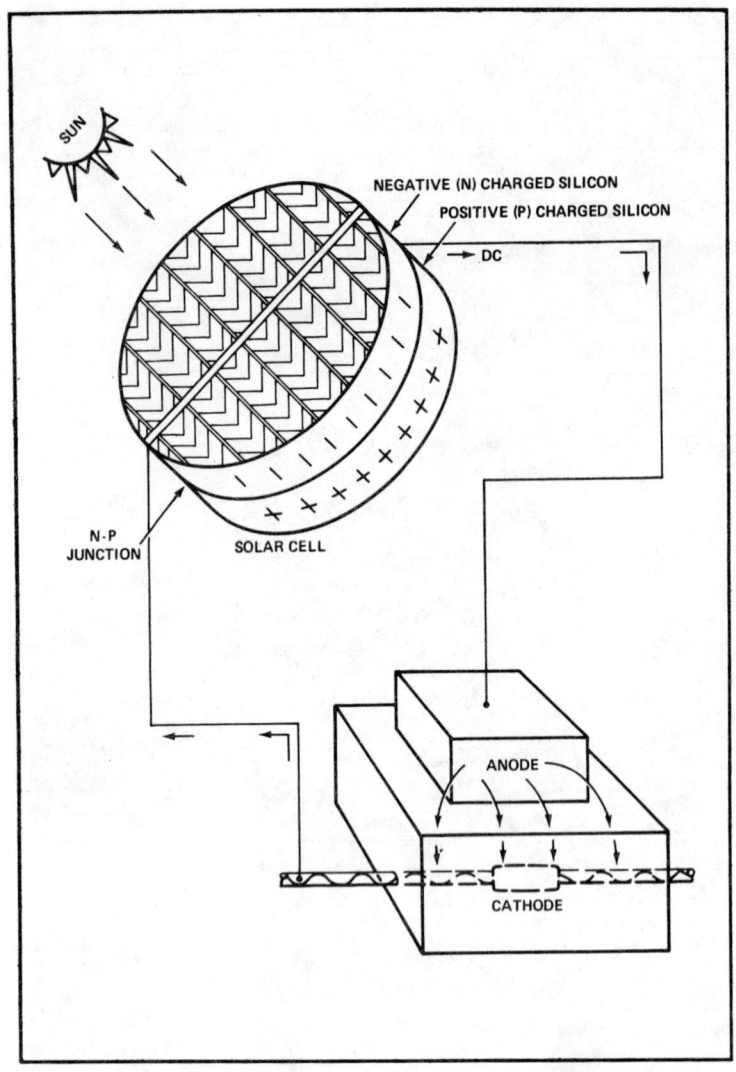

Fig. 5-11. Principle of electrolytic corrosion.

provided for warning light, bells and gates required to protect these crossings day or night. A solar-powered railroad crossing is shown in Fig. 5-14. Track circuits and semaphore signals, which are railroad safety devices that indicate the presence of a train on a certain section of the track, could also be powered by solar cells. In all cases, solar generators have replaced primary battery systems supplementing power for these key safety devices. Another important application is that of railroad communication networks. In many

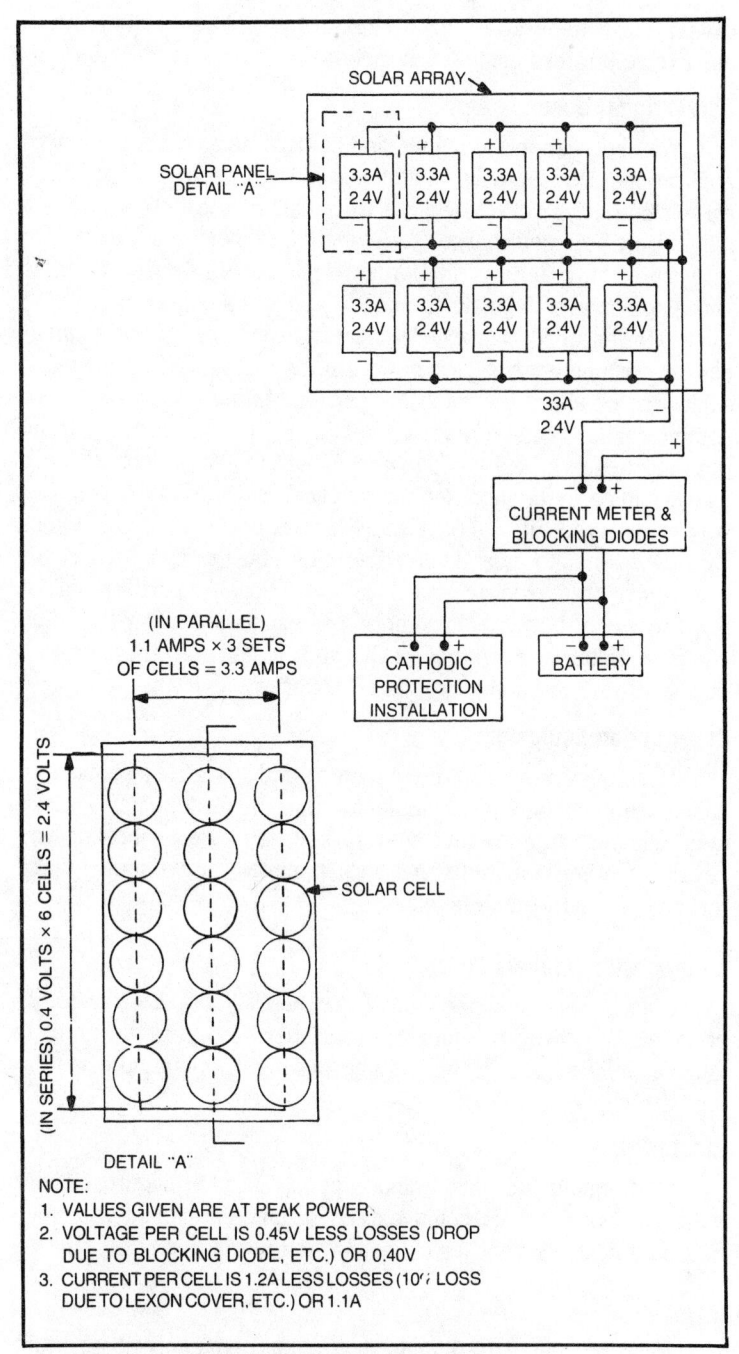

Fig. 5-12. Solar-powered cathodic protection system.

cases, solar generators have replaced the traditional thermal-electric generators using fossil fuels.

Recreational Uses

With recreational activities on the increase today, solar power applications are also daily multiplying every day in this field. Golf carts using solar cells to charge their batteries can now go for a longer time between charges and in many cases extending distances that can be travelled by 100 percent solar-powered. Motor homes with solar arrays eliminate the need to start the engine every time the batteries need charging. Appliances can now be run without danger of running the battery down. Motor boats and sail boats using solar panels to keep batteries charged for lighting and ship-to-shore communications can be ready to go at a moment's notice. Normally the batteries in a boat run down while sitting idle in docks. These panels can easily be mounted on the cabin roof or other out-of-the-way places (see Fig. 5-15). Campsites will not be limited to those with power hookups, and campers will be free to camp wherever they wish if photoelectric generators are used. Backpackers will not need to worry because small solar cell panels are lightweight and compact. Hikers can now have a CB radio, lights, and other conveniences without the added weight of batteries.

Construction Equipment

Dead batteries from frequent starts and stops are a problem to construction industries. Equipment sitting idle for long periods also needs protection and some way to keep batteries charged. The severe weather conditions that solar modules can withstand make them ideal for these situations.

Electric Gate Openers

Solar cells have been used successfully to keep batteries charged, and these openers allow ranchers to operate automatic gate openers to get through without having to open and close the gates themselves.

Electric Cars

With rising fuel costs, more people are turning to electric vehicles. Solar panels could extend distance travelled and keep batteries in these vehicles charged when not in use.

Toys

Probably one of the biggest applications of the immediate future will be using solar cells in toys. It is fun to play in the sun with solar-powered trucks and cars. The toy counter of the future will be filled with solar gadgets, like those in Figs. 5-16 and 5-18.

Watches and Calculators

In this day of electronic watches and calculators, solar cells become even more attractive for use in such applications.

Fig. 5-13. Educational television used in Niger (courtesy Solarex).

Fig. 5-14. Railroad grade crossing in Rex, Georgia (courtesy Solarex).

Miscellaneous Use

Because of the unique characteristics of the solar cells, new applications are being found daily. It would be difficult to mention them all in this book. Almost any electrically run device has the potential of being run on solar power. Solar cells are now an attractive alternative in areas where diesel generators are producing electricity at (more than 20 cents per kilowatt). Remote highway emergency boxes are powered by small solar arrays in many of the national parks. Solar electricity powers many environmental monitoring devices for indicating the amount of water or air pollution in an area. The United States Weather Bureau has many remote meteorological observation stations now using this convenient form of power. A wide range of additional applications includes the following: airport landing lights, weather reporting stations, water pumping, power for second homes, emergency location alarm transmitters, electric fences, intrusion alarms, highway signs, dust warning signs, and many more (see Fig. 5-17).

But the remote-application market alone will not provide growth for industry to sufficiently expand its operation and production techniques. As the price of cells goes down, the many new market areas that begin to open will increase the production level. It is to this end that many of the government-funded projects are aimed. By putting in large photovoltaic generating facilities, the market will be stimulated to higher production.

LARGE SCALE APPLICATIONS

Many of the larger applications to date wouldn't be considered cost effective for a private business, so most of these are experimental units funded by the government. In 1973, one of the largest solar arrays built up to that time was completed for the Sky Lab base station. This 10,000-watt unit supplied the power for the station, which was occupied by a team of astronauts for 84 days.

Solar-cell panels are now employed in such areas as Mead, Nebraska, where they are being used to power pumps for irrigation. This facility will irrigate large areas of corn in summer and then run fans for crop drying in fall and winter.

On an Indian reservation 20 miles south of Albuquerque, New Mexico, photovoltaic modules are being used to replace a 12-foot windmill for pumping water. The system runs a 36-volt, ½-horsepower motor connected to three 12-volt batteries. The pump was designed to use batteries to start the motor, run for 2½ minutes, then switch directly to electrical power provided by the solar panels. If there is sufficient power coming from the panels, it will continue to operate. If it is cloudy and overcast so that the solar panel is not producing enough power, the unit will switch back to the battery, continue to operate for a few minutes, then turn off. In this way, it will try again later, but not run the batteries completely down at any time.

One of the largest solar projects underway at the present is at Mississippi County Community College, which is shown in Fig. 5-19. They are building a new campus which will receive most of its electric energy from solar cells, as well as solar heat for its buildings. The photovoltaics used at this facility will produce about 250-350 peak kilowatts of electricity. This power will be stored in special batteries. It will also be connected to an electric utility grid for back-up power. A computer controller will maximize solar power

Fig. 5-15. Recreational sanitary facility (courtesy Motorola).

Fig. 5-16. Miniature solar power plant kit (courtesy Solar Clover Corporation).

usage. Because of the large number of cells used in this application, they were able to buy the cells at a very competitive price, showing that the price can come down as the quantity of cells being purchased goes up.

The future of solar cells looks great. Soon, they may be providing electricity for central power stations such as the one at the Phoenix Airport, which will supply power to some of the new airport terminals. They may be employed by utilities to meet the growing demand for electrictiy, or they may be used in individual homes or offices. General Electric is under contract by the Department of

Fig. 5-17. Arizona Public Service Company uses Solarex Unipanels to power this radio repeater station despite its proximity to one of its own 500,000-volt lines (courtesy Solarex).

Energy to evaluate and develop photovoltaic systems for residential use.

One of the first of these applications will probably be a partial air-conditioning of shopping centers and office buildings. A significant part of the air-conditioning load in such structures is the solar load. This means that solar-driven air-conditioning systems designed to meet the solar fraction of the load would be available when cooling is needed most; in fact, when it is cloudy or at night, cooling would not be needed at all. In other words, when cool air from the air-conditioning is needed, the sun will be shining the brightest. This will in turn run the photovoltaics to power the air-conditioning units.

Fig. 5-18. Solar-powered propeller kit (courtesy Solar Clover Corporation).

Fig. 5-19. Mississippi County Community College—the world's largest photovoltaic installation (courtesy Solarex).

In southern states, air-conditioning in large businesses and commercial structures is virtually required year-round. Air-conditioning is needed on almost any sunny day throughout the year. The large commercial users of electrical energy normally pay demand charges, which increase in the southern part of the country where electric utility peak loads occur in the summer time. The demand charges paid on the year-round basis are related to the peak summer demand, which inevitably correspond to the hot, sunny summer days. Thus solar installation designed to carry the solar portion of the load would have a very large impact on the demand charges which the user would pay. The solar installation itself would be relatively simple with photovoltaic devices driving the electric compressors and chillers. Since the availability of solar cooling would correlate rather closely with the cooling demand, little need for energy storage would exist, and such storage could be provided by using chilled water storage.

A large number of office buildings and shopping centers in the sun-belt states would benefit by using photovoltaics for this type of application in the next five years. It is still too early to try to depict the precise timing and development for this and other early photovoltaic markets.

Designing A Photovoltaic System

6

In order to take advantage of the unique features of photo cells, careful consideration must go into the design of the system. The photovoltaics themselves have their own strengths and weaknesses. When these are multiplied several times, they become important considerations. The way in which the system will be used will make a great difference in its performance. The amount of sunshine available in a particular location and the weather conditions that exist will all affect the size of the system. Load requirements, the time of day when the power will be used, temperatures, and unusual weather patterns must all be considered in the design. Estimating how long the cells will last is important in determining their cost. Where the system will be mounted and how it will be protected and maintained are just two of the factors that must be contemplated in designing a photovoltaic generating system. In this chapter, we will be discussing these important design considerations in more detail.

COLLECTING DESIGN DATA

When designing a solar array for a reliable solar power system, many factors must be considered. The seasonal variations must be taken into account. The total number of hours that the sun will be shining daily as well as total hours per year must be calculated. This will be affected by the weather conditions and changing angle of the sun from summer to winter. Many local considerations, such as the altitude of the site, mountains, lakes, and cloud cover, all affect the amount of sunshine which will directly or indirectly reach the solar array. Even though there are many variations in geographical locations, solar energy will work almost anywhere in the United States. In less desirable areas more solar arrays or cells will be required. Solar cells are rated in the amount of peak power they will deliver at noon on a clear day.

One of the first steps in designing a system is to collect all the information needed to give a proper perspective. Before any decision can be made, you must have an idea what you will be working with. You must gather specific information about the amount of sunshine received in your area. You also need to know what weather variations exist which might affect your solar collection. Most of this information can be obtained from your local weather station or by writing the National Weather Record Center in Asheville, North Carolina.

Detailed information is needed concerning the amount of power you will require. You will also need to know how the power will be used during the day. So much information is needed in order that the optimum system may be designed. With a properly sized system installed, virtually any power requirements can be obtained. Let's take these areas one at a time and determine the best way to collect the design data.

Radiation

The radiation of the sun (the amount of sunlight striking the surface of the earth) varies from one location to another throughout the earth. The amount of sunlight falling on a certain location is affected by many factors. Although the amount of available daily sunshine varies with these environmental factors, it is still possible to determine the average solar radiation for a given area. See the solar insolation maps in Figs. 6-1 and 6-2. The amount of the sun's energy received each year is very constant. From one year to the next, the variations are less than 10 percent. Seasonal and daily weather conditions vary greatly and are often unpredictable. Solving these extreme differences has been one of the challenges facing the designer of solar systems. But weather-data centers have done a lot in recording and averaging the amount of sunshine in various localities over periods of several years. By contacting these weather centers and referring to this data, a person can get a good idea of the amount of sunshine he has to work with each month of the year.

Weather Patterns

By averaging the amount of sunshine falling on the earth over a several-year period, fairly accurate data can be obtained. Weather

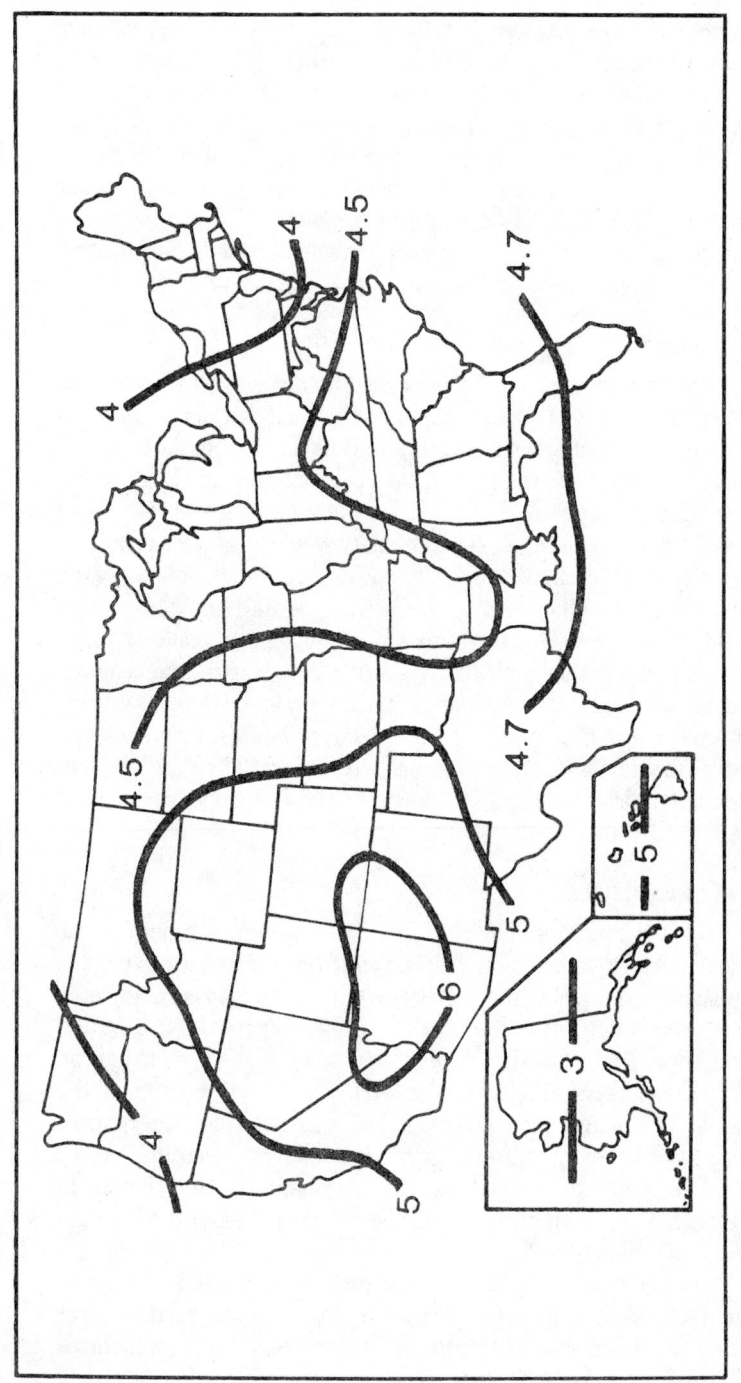

Fig. 6-1. Yearly average peak sun hours per day (courtesy Solarex).

conditions, however, are a little less predictable. The weather seems to change in certain cycles over a number of years. Accurate records of the amount of cloud cover, rainfall, and prevailing winds can be obtained by consulting the local weather station. These records of the weather conditions over a long period of time prove helpful in adjusting the amount of sunshine expected, and the amount of storage needed if a 24-hour system is desired. There is no way to change the weather, so these conditions must be taken into account when designing a solar system.

Sun's Angle

Because the incident angle of the sun changes throughout the day, month, and year, the output of a solar module will vary greatly. When the solar panel is parallel with the sun's angle of incidence, the greatest amount of energy can be received. As this angle is increased, the expected power will vary at a geometrical rate. On a fixed module, the greatest solar intensity would be at noon on a clear day. It seems that the air mass also has an effect on the output of the cells. Because of air mass between the sun and the collectors on earth, the amount of solar radiation striking the module will also affect the production of electricity. A study by General Electric will analyze various climatic regions in the country and will develop ways to predict the performance of photo-electric systems. From this, the cost of residential systems can be determined. They are also developing detailed designs for single-family dwellings suitable for photovoltaics by the mid-1980s.

Power Requirements

Each system designed will have different power requirements or loads. Much depends on the load and how and when it is used in determining the size of the system. Whether the load is steady or in surges might determine the type of storage required. If the system uses power during only the daylight hours, a storage system may be eliminated entirely. But before the total system can be designed, the amount of voltage and amperage required must be determined. Also, the frequency of use must be decided to evaluate the power loads. At best, the loads must be averaged, so it is better to overdesign the system to handle the occasions when the load exceeds the average.

To get some idea of the load demand expected during the course of a single day, refer to Fig. 6-3. The daily and annual energy cycles can be compared to compute the average power essentials (Fig. 6-4). The average power can be used in determining the size of

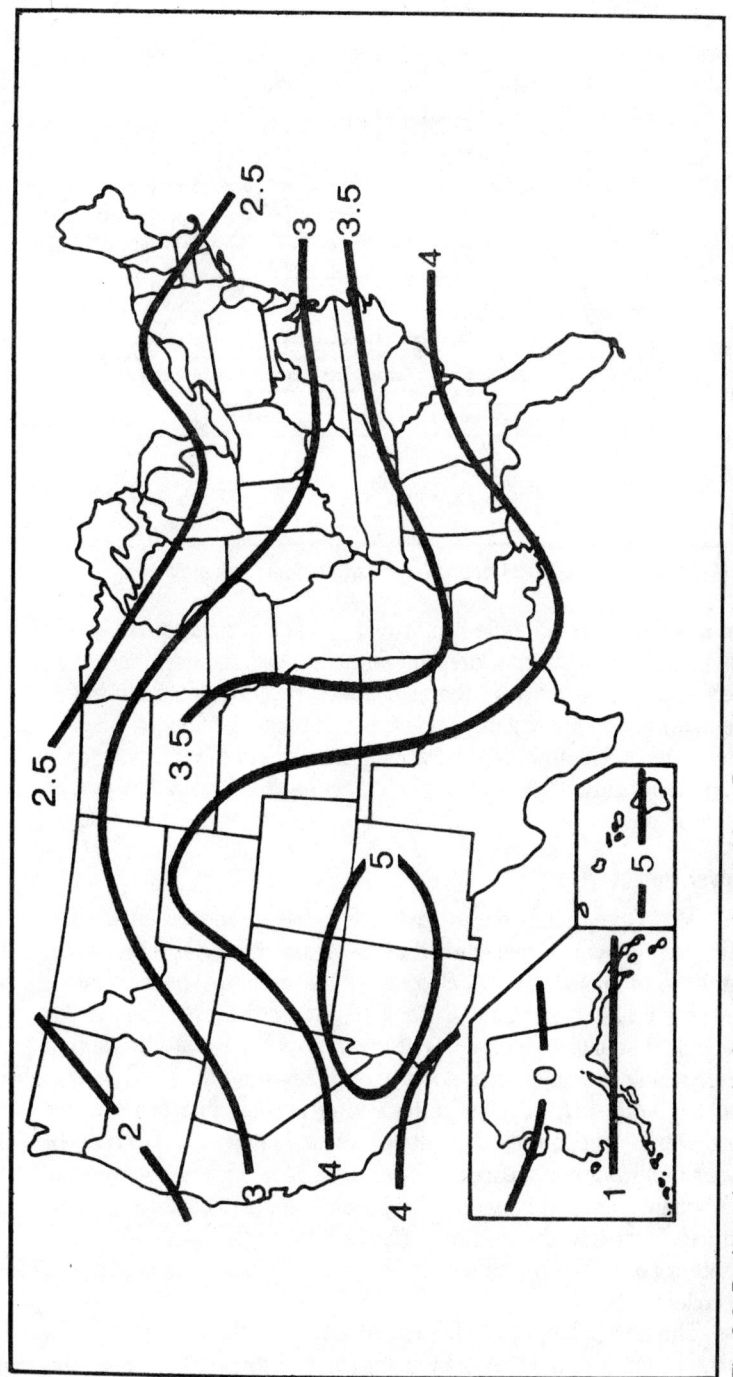

Fig. 6-2. Peak sun hours per day for a four-week period from December 7th to January 4th (courtesy Solarex).

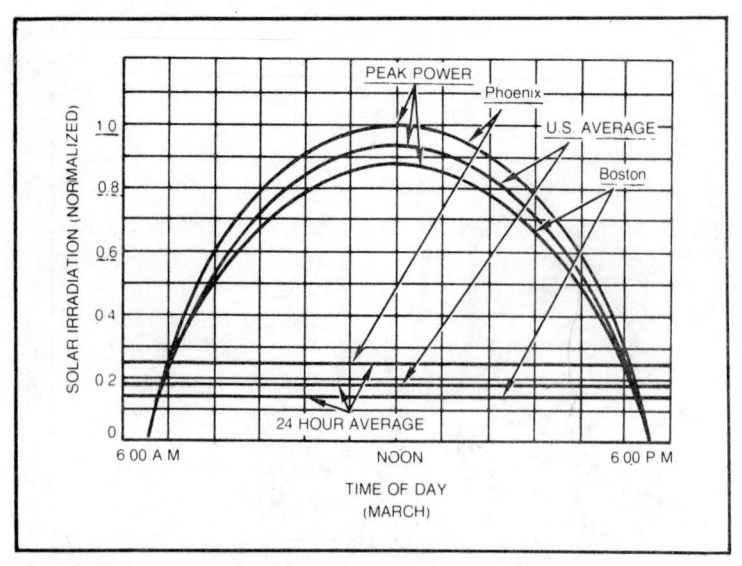

Fig. 6-3. Ratio of peak to 24-hour average power (courtesy Motorola).

the storage system needed to provide power during periods when the array output cannot keep up with the load requirements. So, then, the amount of power that is needed, the geographical location, the average amount of sunlight per year and other considerations will affect the solar-power generation. The environment, including weather, pollution and seasonal changes, will influence the power requirements.

Power Output

After the design data has been gathered and received, the necessary power output should be determined. Other information gathered at this point will also affect the output. The amount of power which the solar panel will deliver is directly proportional to the amount of sunlight received. The most important consideration in determining the amount of solar energy reaching the earth is that sunlight is *diffused*. To convert this energy into a usable form here on earth, large collector areas must be exposed to the sun. We receive about 1 kilowatt per square meter or the equivalent of 1 horsepower per square meter at noon on a bright day. We also have to contend with the fact that the sun only shines during the day; even then, weather conditions may affect the amount of sunlight we can expect to collect.

The power output can be determined by taking all these things into account. The number of modules can be estimated to satisfy the load demand. To do this, the number of langleys (see Figs. 6-1 and 6-2) must be converted into sun hours. The optimum tilt of the array for year-round collection should be about 45 degrees in most areas of the United States. And, of course, the panels need to be facing a southerly direction. It is also assumed that the panels will be mounted where they will not be shaded during the daylight hours. To determine the number of modules needed for a certain application, the following equation can be used. Average daily load in amperes per hour at operating voltages plus 20 percent for system losses over peak sun hours at the site equals the total output required in amperes. By using this formula, the power output can be obtained from information collected about the sun's radiation and the load requirements.

$$\frac{\text{Average Daily Load in Amp. Hours} + 20\% \text{ System Loss}}{\text{Peak Sun Hours}}$$

$$= \text{Total output in amperes}$$

The total energy output that could be expected per square foot is shown in the chart in Fig. 6-3. As displayed, the difference between the best and least desirable locations can vary significantly.

Sizing the System

In sizing the system, much is required in gathering data to obtain the optimum design. The object is to get the best system for the least amount of money. To do this, a good combination of solar modules and batteries must be found. After evaluating the design

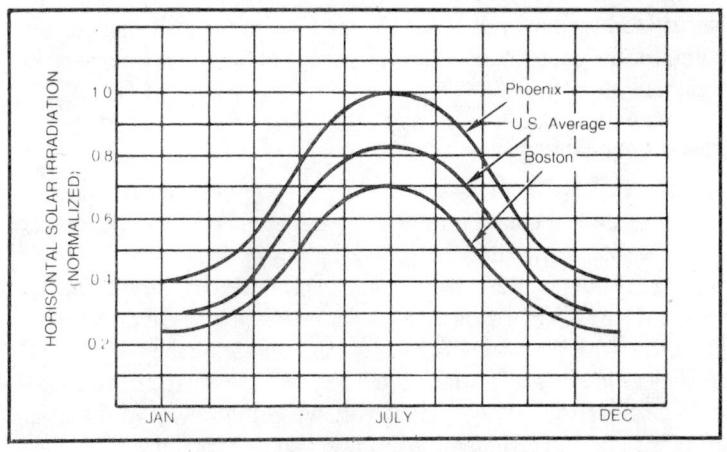

Fig. 6-4. Annual horizontal distribution of solar energy (courtesy Motorola).

data, the smallest size array possible that can handle the load is determined for each day of the year. Based on the insolation charts, the output is then compared to the expected load demand for each day of the year. The difference is then multiplied by the number of days when the panel output falls below the load. This indicates the number of days of storage needed to ensure continuous operation. The number of days of storage is determined for each of the different array sizes, and the most economical system is then selected. To assure adequate supply of electricity when it is wanted, the system must be designed to cover times of cloudiness, rainy days, and nighttime use.

Selecting Batteries

The excess electrical output can be stored in batteries until needed. The designer must also take into account the battery discharge rate. Batteries have a maximum discharge rate and also a maximum rate at which they can be recharged. Each battery has a self-discharged rate which should be considered along with the life expectancy, which would affect replacement costs.

The amount of power is directly proportional to the size of the system. As the total system is laid out, designers will need to be aware of how to control and transfer the energy. Insulators, voltage regulators and storage batteries will be needed for a complete photovoltaic system. As the solar array converts the sun's energy into usable DC power, it then flows from the array to the storage battery as shown in Fig. 6-5. As it does so, a voltage regulator controls the degree of energy flowing to the batteries much as the voltage regulator in a car keeps the battery from overcharging. The battery can then provide power in times when the load exceeds the output of the solar array or during periods of darkness or inclement weather.

Power Conditioning

Since most of today's electrical applications use AC power, additional power converters are needed to change the DC power from the solar array into an AC power. Many options are available for storing the power after it has once been produced by the solar array. One of the best ways of using the solar power is to use it directly in a DC form. With the advent of recreational vehicles, many 12-volt DC appliances are available which could be used directly from the photovoltaic or battery storage. However, when the power of the solar array is to be converted into AC power, some efficiency will be lost (see Fig. 6-6). By making this conversion, however, the power can

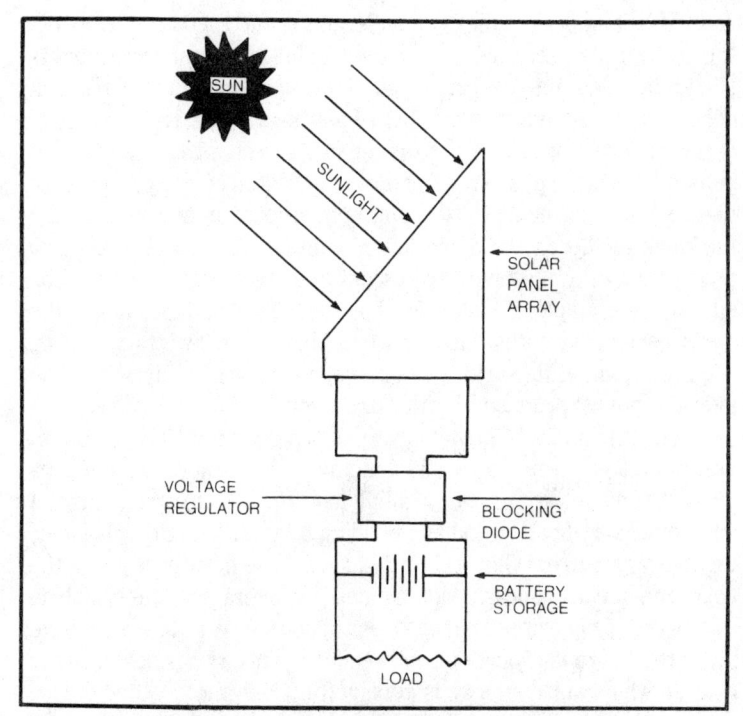

Fig. 6-5. The flow of electricity from sunlight to the storage battery (courtesy Solarex).

be interphased with the public utility's power, which can be used as a storage system. Before installing such a system, however, this must be carefully checked through the local utility company to make sure this type of an arrangement will be satisfactory.

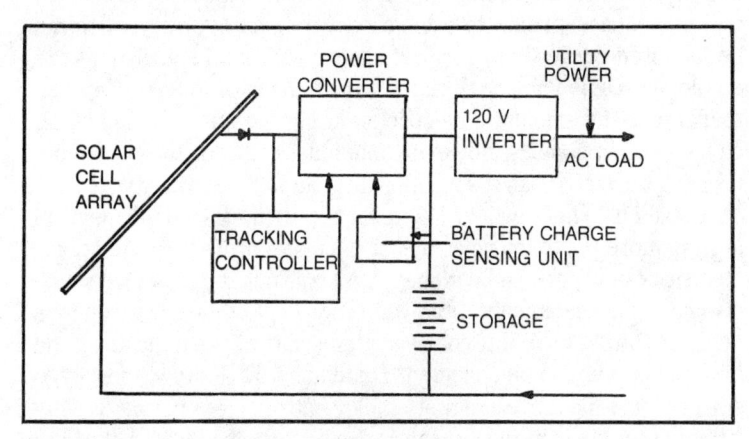

Fig. 6-6. AC inverter and the public utility power.

After taking all of these considerations into account, the cost of the system can then be determined. The load required in watts should be taken into account when figuring the cost and number of solar arrays needed. The number of watts as indicated on the solar array must be multiplied five to seven times to give you an equivalent peak watts. Since peak watts of power indicated on the photovoltaic device are determined by the optimum amount of sunshine during the peak of the day, a much lower amount of wattage is actually produced by the photovoltaic system over the entire period of the day. As an example, if 1 kilowatt of power is required throughout the day's period, a photovoltaic array of about 7 milliwatts would be required to meet this load. Peak power rating is much less than the average power produced by the photovoltaic device.

The designers of power systems typically rate the power required for a given application by defining the current drawn from the system at a specific voltage. Multiplying the total amount of current used over a 24-hour period and dividing by the time intervals yields the average current that will be required. The power output of the solar module is similarly determined by averaging the sunshine cycles on a daily basis. These cycles, of course, will vary considerably with geographic locations and seasons. This data is based on the rate at which solar energy is received by the photovoltaic device during the average sunlight hours per day. A chart showing the daily distribution of solar energy is shown in Fig. 6-3. The photovoltaic current is produced at a rate proportional to the amount of sun shining on the solar device.

To receive the maximum efficiency from the solar panels, they must be mounted permanently at a 45 degree angle, oriented south and permanently exposed to the sun. But other methods of installation may be acceptable. In some cases, the photovoltaic device can be mounted vertically or in other cases, horizontal installations are employed. Depending on the situation, tracking devices can be used to increase the amount of electricity during the entire period of the day. In such instances, however, one must consider the cost of the tracking device versus the additional amount of electricity which can be obtained. There would seem to be, then, a tradeoff point in getting more photovoltaic arrays or panels to provide the additional electricity or in putting the money into a tracking device which would also increase the amount of power. There are, then, many options open for the user of photovoltaic systems in the size of the array and how the power can be collected. Depending on the type of storage selected, the system can either employ batteries or operate directly without storage. The systems may also be interphased with the local utility power systems.

OPERATING TEMPERATURES

Solar panels operate in a wide temperature range between a minus 70 degrees Fahrenheit and a plus 150 degrees Fahrenheit. However, the amount of power coming from the panel varies somewhat with the temperature. As the panel temperatures increase, the voltage decreases accordingly, while the current seems to remain relatively constant. At lower temperatures, the power increases with drop in temperature, and again, the current remains almost the same. The solar panels can operate in a wide range of temperature without changes in specifications. Standard tests measuring the influence of temperature on electrical output show that at 25 degrees Celsius the voltage will vary inversely with the temperature at about .2 volts per degree Celsius. But the current changes about 0.3 percent degrees Celsius due to the compensating effect of the current with the temperature. Changes in ambient temperature affect the solar array and also the voltage of the battery. As the temperature increases, the maximum voltage output of both the solar array and the charging voltage of the lead-acid batteries both decrease. The temperature coefficients of the solar cell is a minus 2.2 millivolts per degree Celsius. Thus, as the ambient temperature increases, the open-circuit output voltage of the solar array decreases more rapidly than the battery-charging voltage. Most systems, then, are designed for particular maximum ambient temperatures, for example 40 degree, and thus are capable of operating in all

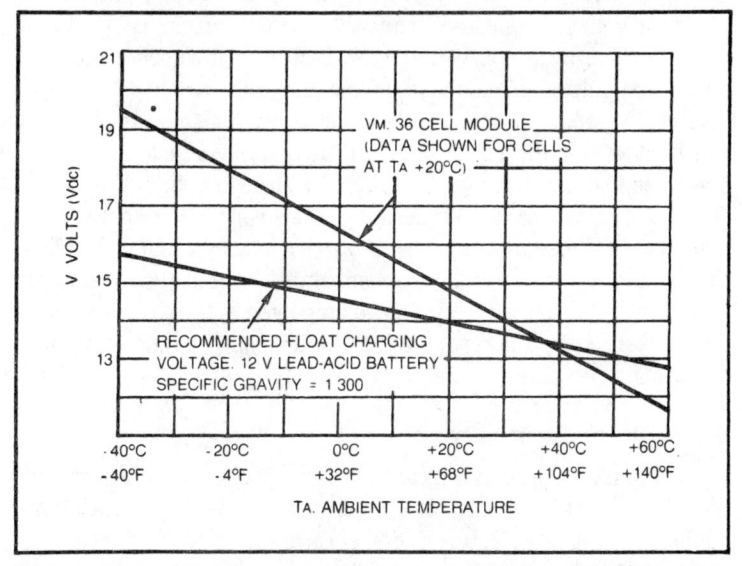

Fig. 6-7. Array and battery temperature characteristics (courtesy Motorola).

ambients less than the specific maximum. The typical battery temperature characteristics for a 12-volt system are shown in Fig. 6-7.

Because of the adverse effects of temperature fluctuations on the solar panels, it is best to situate them where the temperature range will not vary significantly. In some cases, the photovoltaic panels are made to have air circulation behind the cells, carrying off the unwanted heat. Other systems have been designed, particularly when using a concentrator device, to circulate water behind the cell-mounting platform and carry away excessive heat. In such a device, the heat can also be used for thermal potential. In some cases, this has been used for domestic hot water and also has been employed in space heating. So the excess heat buildup in the solar module can be advantageously used in other ways. For maximum cell performance, however, some effort should be employed to keep the ambient temperature within the maximum cell-production range.

Batteries

Various types of batteries are used with photovoltaic systems. In all cases, protect the batteries from extreme temperature changes. Some of the batteries most commonly used are rechargeable *nickel-cadmium* batteries, rechargeable *oxide* batteries, and the nonrechargeable *silver-oxide* batteries. The nickel-cadmium type seems to be less sensitive to overcharging, and is the type of battery being used in watches and calculators. The oxide battery seems to be very sensitive to overcharging and should be used only in systems having a voltage regulator. The silver-oxide battery is generally considered to be a nonrechargeable battery. However, if care is taken in supplying proper voltage and amperage to meet the manufacturer's specifications, this type battery can also be recharged. But in all cases, the temperature of the battery affects its overall operation, so care should be taken to place the batteries in an isolated place where the temperature range will not vary greatly. In many cases, these batteries are located in a small room or building that is insulated form the outside ambient temperatures. In smaller installations, an insulated box can be employed to house the battery. The battery should be protected to assure long life and maximum performance in all situations.

Solar Cell Life Expectancy

Because the silicon cells are similar to glass, their life is almost unlimited when kept at room temperatures. One thing which may hamper the cell is the deterioration of the grid pattern on the surface of the cell. The soldered contact points which interconnect the cells

Fig. 6-8. Solarex solar electric system powering a marker beacon for an airport in Kenai, Alaska, for the Federal Aviation Administration (courtesy Solarex).

in module design may also become defective. These contact points and the grid patterns that are used on solar cells are both subject to corrosion. To prevent this corrosion, most cells are built into a panel with a coating material completely covering the cell. This coating material must be highly transparent and weather resistant to protect the cell from adverse weather conditions. Therefore, if properly protected, the solar cells should last indefinitely.

Much work is being done to provide maximum life expectancy from the module design. Where many cells are connected together to create the module, there are many contact points which could be potential problems. If properly connected and protected from the environment, however, these modules should show no signs of deterioration. Another place where the modules may give some problems is connecting the modules into solar arrays. These connections, however, should be made behind the solar modules to protect them from evironmental conditions and corrosion of the contact points. If care is taken, the connecting solar arrays should not offer any particular problems.

Along with the extreme care taken when making a solar cell, the next important step in the module construction is that of the cell encapsulation. After the modules have been assembled and all the contact points made, tests are conducted to assure the continuity of the system. Several methods of encapsulation have been employed by the various companies producing solar cells. One of the most popular methods has been that of using silicon rubber injected around the cell, completely surrounding its surface area. The silicon rubber, however, is an extremely sticky substance, and thus has a high dust collection capability. Consequently in most applications, glass is preferred as an outer covering to protect the cells, with the silicon injected behind the cell and between the glass front and the rear surface of the module. If care is taken to make sure that all of the air pockets and bubbles are out of the silicon material when the modules are made, there should be no way contaminants could get into the cell to deteriorate its electrodes and contact points.

When installing a solar system, much care must be taken at the various points to assure system reliability. The carefully produced solar module must be attached to some type of supporting device to protect it from wind and other adverse weather conditions. When the system is wired properly, it should be protected from sun, wind and rain (see Fig. 6-8). Photovoltaic systems can be mounted on the roof of a building or mounted on an isolated structure. In either case, the system should be placed fairly close to the battery storage and load. The amount of line loss in transmitting the power through long

Fig. 6-9. Solar cells power this military range instrumentation system (courtesy Solarex).

Fig. 6-10. A solar array provides power for pumping water in Mali, West Africa (courtesy Solarex).

power lines is thus eliminated. This is especially true when using DC power.

One of the chief advantages of photovoltaic systems is the low maintenance required to keep them in operation. Unlike most other systems, the solar cell has no moving parts. Because of this, many of the mechanical problems associated with other power generators are eliminated. Solar electric power is produced, making no noise and creating no pollutants as a by-product. It offers a clean, reliable source of power with high life expectancy. Photovoltaics, then, can be an extremely reliable system, supplying power with almost no attention needed, such as in those areas shown in Figs. 6-9 and 6-10. With all these potentials, solar electrical power is bound to have a definite impact on our electrical generation in the near future.

Cell Materials and Their Development

7

It seems that in the quest to produce a better and more efficient solar cell, many types of materials have been employed. *Selenium*, one of the first materials used in solar cells, has very low efficiencies. But because of its low cost and ease of production, it has become successfully used in photography for light meters. As researchers and developers of the transistor searched for materials to be used in that industry, they discovered that *silicon* had properties which would provide a voltage when exposed to light (see Figs. 7-1 through 7-3). Silicon has received a great deal of interest because it is one of the most common materials found on earth. The silicon cell is many times more efficient as a solar device than the selenium cell. Because of this relatively high efficiency, silicon has become one of the leading materials used in the photovoltaic industry.

Solar cells may be divided into three categories: space cells for use in outer space; terrestrial or earth-type cells; and photosensors used in light-sensitive and photographic equipment. Let's examine each of these three areas to see how they differ.

SPACE CELLS

Millions of solar cells have been produced, most of them used in space satellites. In fact, the first major use of solar cells was in the space program. The main ideas in this usage were efficiency and compactness. These cells were of the highest quality and efficiency in order to meet the standards of the space program. Space cells had very thin grids and were very delicate. The solar cells made for the

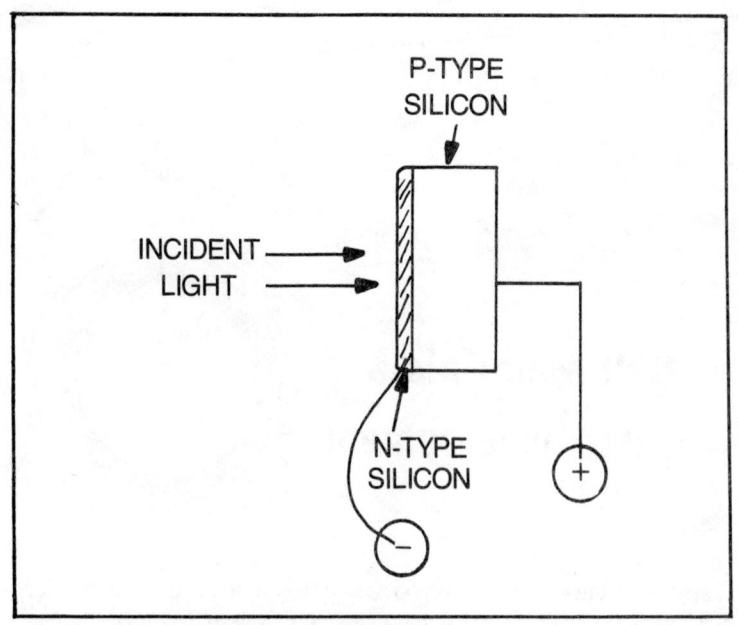

Fig. 7-1. Elementary silicon photovoltaic cell.

space program were very expensive, but it was of little concern since the photovoltaic array was such a small part of the total cost. The key is *watts per square centimeter*, or *watts per gram* of weight. There were few alternatives because conventional batteries would soon run down and become useless with no way to replace them in outer space. The space cells were square or rectangular because they needed to be placed close together with very little wasted space in between. They were a dark blue color, having been coated with an oxide to reduce optical reflection.

The need for a more efficient power source in the space industry led scientists to use silicon as a base material for the manufacturing of photovoltaic cells. Much higher efficiencies were reached with this new material. For many years, this material was used almost exclusively in the space program. But as people began to show interest in photovoltaic applications on earth, the price became a very limiting factor. In order to make the solar cells more attractive, something had to be done to get the price down or to improve the overall efficiency.

TERRESTRIAL CELLS

At first, many of the cells used in the terrestrial applications were rejected solar cells from the space industry. Not being as concerned about the wasted areas between the cells as in the space industry, terrestrial cells are generally circular discs, such as shown in Fig. 7-4, or segments of discs. Because of the manner in which solar cells are produced, it is more economical to leave the cell in the original round shape when cutting them from circular ingots. The terrestrial cells also are different from the space cells in that they have larger grid patterns. These cells are made larger to increase the overall efficiency and reduce the amount of material needed to produce the cells. Because the amount of power generated is proportional to the area, it is important that the solar cells be as large as possible. Some of these cells may provide as much as two amperes in a single cell. The cost of the cell is expressed in *dollars per watt*, which is important to know when shopping for cells or systems. When compactness is desired in terrestrial applications, hexagon-shaped cells reduce the waste areas between the cells, and they do not require excessive trimming. Thus, this method achieves high density of cells in large groupings.

Solar cells with special responses are available for particular needs. The cells may be purchased with color-coded plus or minus wires ready to be installed in panels. Cell efficiencies have been improved by applying a transparent protective coating with antireflective qualities. The cells are also made in two types: an NP type, which has the negative terminal being on the dark side of the cell; and the PN type cell, which has the positive terminal on the light side and a somewhat higher efficiency rating.

PHOTOSENSORS

Special solar cells have been produced for photographic light meters and light-sensitive equipment. These cells must be of high quality. Photo-sensitive equipment and pyrometers are used in measuring the amount of solar intensity striking the earth at any

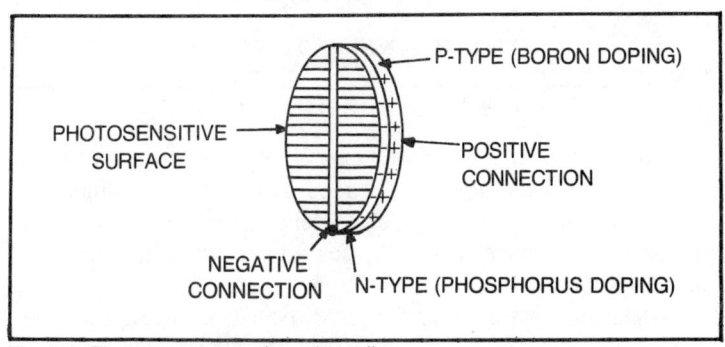

Fig. 7-2. The construction of a solar cell.

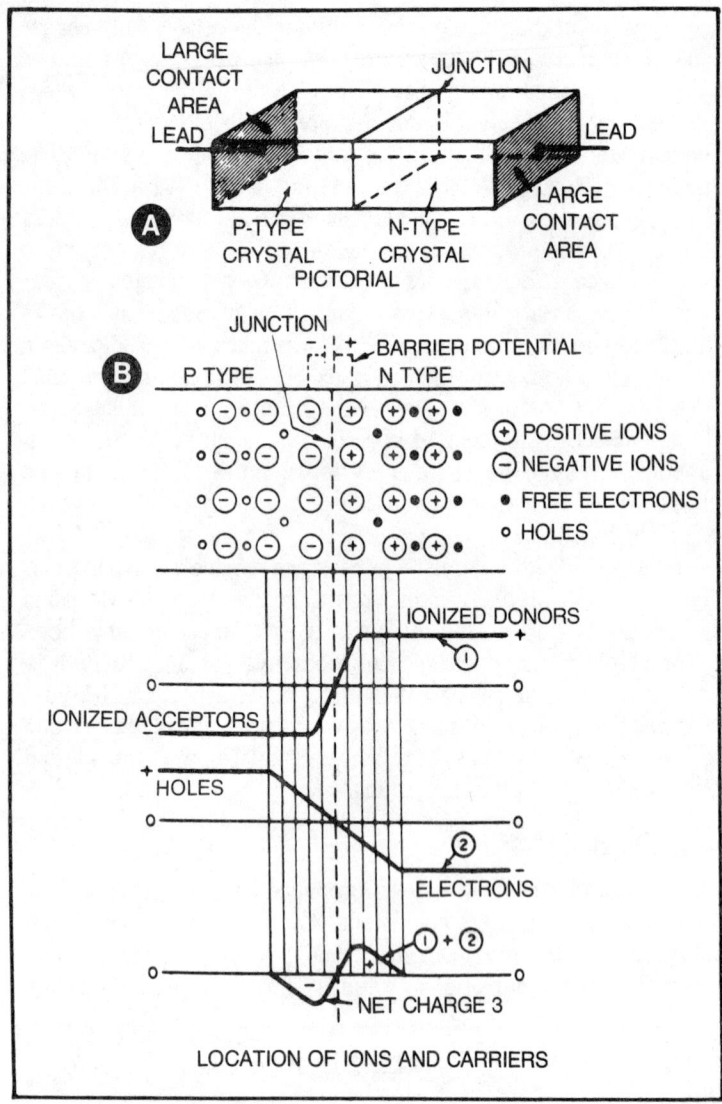

Fig. 7-3. Formation of a PN junction.

given time. Because such meters are so sensitive to the amount of light being received, they are also used in measuring the amount of smoke or smog. A reading can be taken of the density of the pollutants when smoke or smog passes between the light source and the cell.

While the present manufacturing techniques were being developed and applications were employed, the demand for solar cells increased. In 1975, the manufacture of terrestrial solar cells began to exceed the manufacture of cells for the space industry. The rectangular-shaped cells for the space industry were also made from round ingots, but were trimmed into the square or rectangular shapes. The results of these trimmings were mostly waste and thus increased the cost per cell. So to reduce the cost of solar cells, every effort was made to reduce waste and increase the overall efficiency of the cell.

CURRENT TECHNOLOGY

We will briefly repeat how the solar cell is made in current technology and then in a following section describe some of the ways

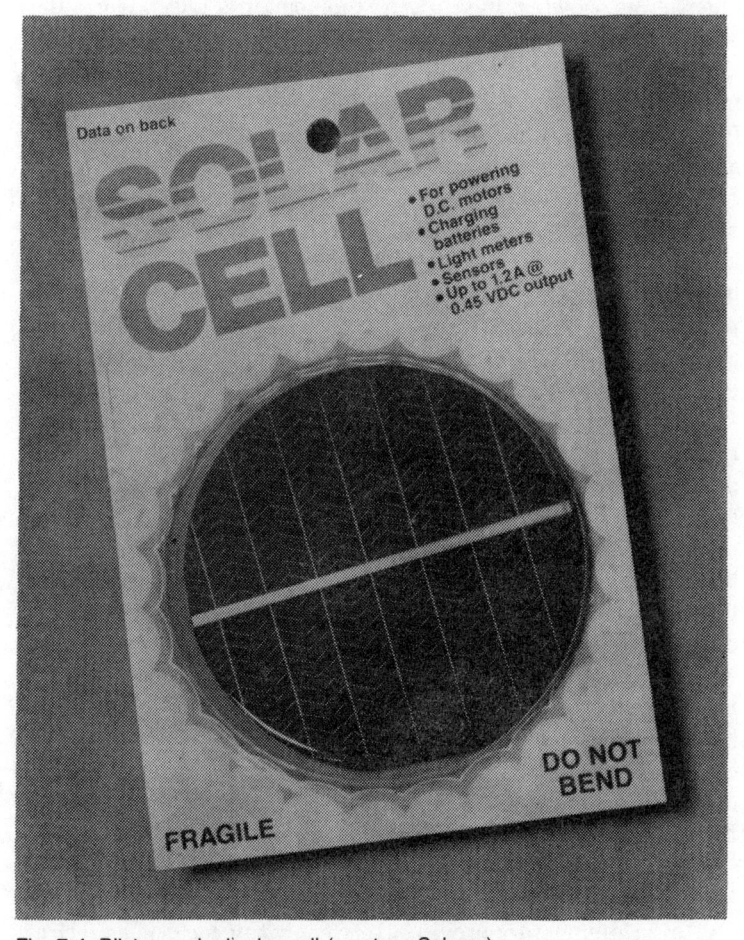

Fig. 7-4. Blister packed solar cell (courtesy Solarex).

of improving the manufacturing techniques to reduce the overall cost of the cell.

The pure silicon is placed into a crucible and heated to become a molten liquid. After the liquid is heated, a pencil-like device suspended over the crucible is lowered, containing a seed crystal in the tip of the pointer. This pointer is lowered until it reaches the top of the molten liquid, at which time, the liquid forms a wicking action and extends up to the pointer. When the seed crystal is heated by the molten liquid, it begins to grow into a crystalline structure, but if allowed to grow in a natural, uncontrolled circumstance, this crystal would set up a random pattern of growth like that of most crystals. Because a controlled pattern of growth is needed to obtain the results desired, the crystal is pulled or lifted slowly out of the crucible, as shown in Fig. 7-5. As the pointer is raised out of the liquid in the crucible, the material which is attached to the pointer becomes cool and hardens. The size of the crystals can be determined by the rate of pull or the speed at which the pointer is lifted from the crucible. If smaller cells are desired, they would be pulled at a faster rate. The larger cells would be pulled at a slower rate. As the crystals cools and hardens, the pulling action makes them all grow in the same direction. These crystals which are pulled from the crucible are called *ingots*. The ingot will stop growing when all of the material is used. However, a certain amount of waste occurs in this process, due to the fact that some of the materials stick to the sides and the bottom of the crucible. There is also a considerable amount of contamination in the silicon which remains in the bottom. So as the silicon is being pulled, many of the impurities fall to the bottom. This reaction purifies the silicon in the ingot. In the early stages of development, scraps, waste or broken wafers are saved and put back in the crucible to be remelted.

Once the ingot has had sufficient time to cool, it is then put on a lathe and milled to the desired diameter. Ingots are then sliced into thin wafers of about 15 to 20 mils in thickness. One of the current problems resulting in the high cost of the cell is in slicing these wafers. The diamond saw used in slicing cells is about eight to twelve mils in thickness, so approximately 40 percent to 50 percent of the ingot is lost in sawdust. This sawdust is unretrievable because it is contaminated by the oil required in the cutting process. Finding more efficient methods of cutting the cells is one of the ways in which the cell prices could be reduced.

IMPROVED MANUFACTURING TECHNIQUES

In an attempt to improve cell efficiency, new techniques such as the following are being explored: first, automation; second, waste reduction; third, ribbon growth; fourth, grid design changes; fifth, multiple cell use; and sixth, thin film techniques. A more in-depth discussion can be made by taking these one at a time.

Automation

In order to reduce the cost of solar cells, much of the labor-intensive process used in present manufacturing techniques must be

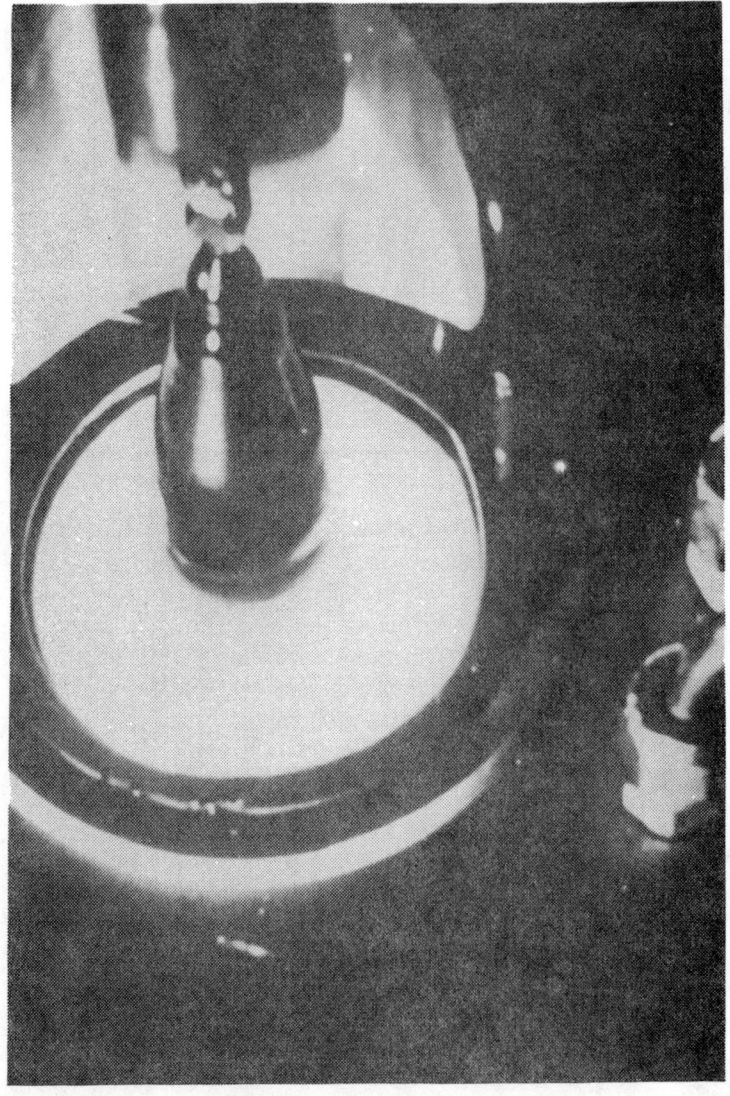

Fig. 7-5. Pulling of ingot (courtesy Motorola).

115

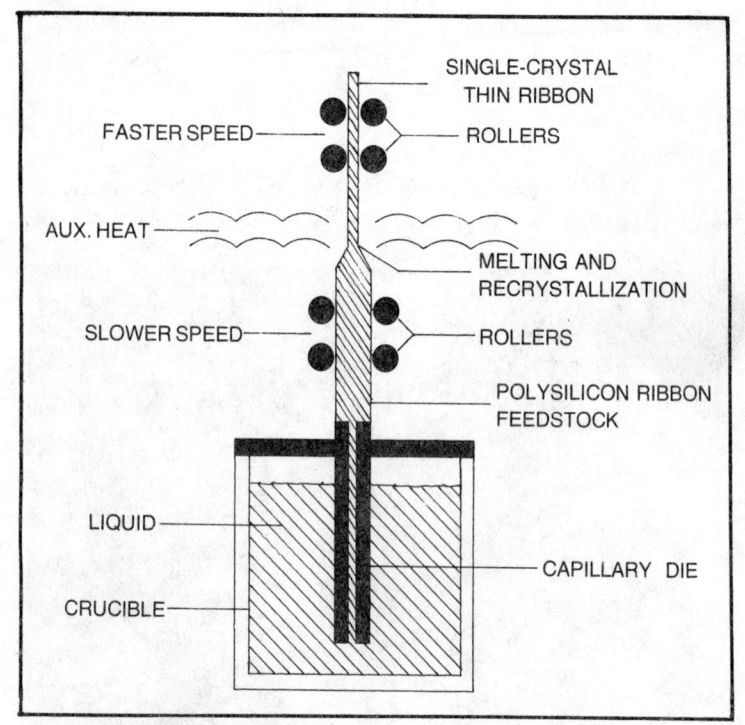

Fig. 7-6. Production of ribbons.

automated. To do this would require very large sums of money. At present, most of the companies are not producing enough solar cells to justify this expenditure. The need for automation is apparent, and time will create the increased demand necessary to bring about this automation. At present, much of the work is done by hand, involving many hours of labor to complete each cell. Handwork is very expensive. In the past, automating a manufacturing process has generally lowered the cost of production many times. We hope that the same will hold true for photovoltaic production.

Waste Reduction

Much of the cost of silicon cells is a result of waste. There is waste in forming the ingots in the first place. A great deal of waste occurs in sawing and grinding the cells to the desired thickness. Much waste results when cells break during the manufacturing process. One of the ways to diminish waste is a new wire saw which will decrease the thickness of the cut to about 4 mils. This would reduce the waste of the cutting process to one-third or one-fourth of that produced by the saws now being used. Since the cost of the silicon material is one of the main costs of the cell itself, elimination of waste would greatly lessen the cost of producing the cells.

Growing Ribbons

Growing ribbons of silicon, a process in which the silicon is pulled through a die, could reduce waste and be simple to automate (see Fig. 7-6). These cells could be produced to reach as high as 10 percent efficiencies. To this date, however, much of the production of these cells has produced poor quality. It is expected to require several years before this process can be improved. Another process currently being explored by Westinghouse under a grant from NASA is a process called *web dendrite*. This process involves pulling thin silicon crystals directly from molten silicon. It is estimated that the price of the silicon cell could be reduced to 75 cents per watt just by improving the conventional silicon slicing and growing techniques.

Grid Design

A most important aspect in grid design is getting the grid elements close enough together to receive a maximum amount of current from the cell with a minimum amount of shading, a process which requires difficult and intricate designs. Most of the solar cells have a large trunkline to which all of the grid patterns tie much like

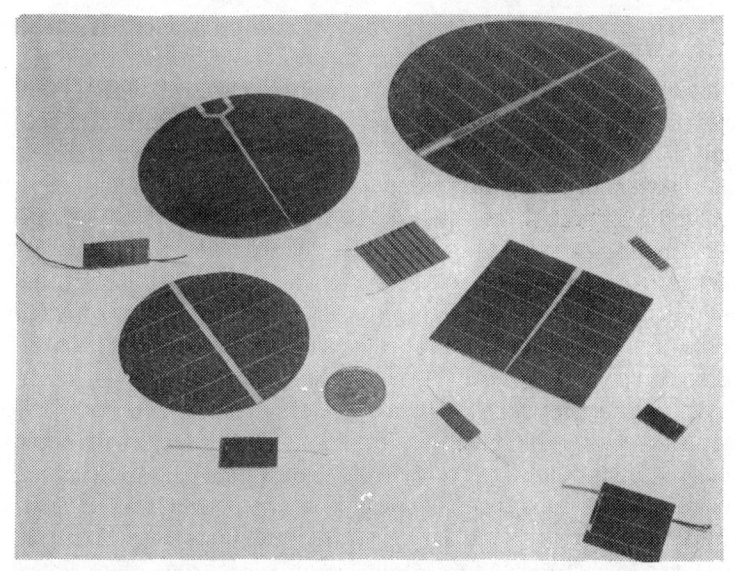

Fig. 7-7. Solar cells range from the four-inch-diameter cell (upper right) which generates over a watt to tiny multicell Microgenerators used to power watches and calculators (courtesy Solarex).

Fig. 7-8. A mountaintop radio repeater (courtesy SES, Incorporated).

tributaries of a river. The reason for the large trunkline in the middle is to carry more current.

Cell Shapes

The many companies involved in solar cell production have tried various cell shapes and sizes, from squares and rectangles to circles and hexagons (see Fig. 7-7). The purpose in the cell design is to achieve the maximum amount of solar collecting area with the minimum amount of holes or unfilled portions in the collector array, as shown in Fig. 7-8. Motorola has developed a new cell which has six-fold symmetry in the metallic pattern as shown in the Fig. 7-9. This pattern minimizes the area of the cell shadowed by metal. The hexagon modules produced by Motorola allow for bonding to be conveniently located where they can easily be connected to adjacent cells. The six metal trunk lines also enhance reliability by providing considerably more connecting points, thus reducing the possibility of short circuits in or between the cells. If one contact of the hexagon-shaped cell is open or broken, the loss in the output of the total module is less than 3 percent. With a cell having only one contact point, a break could result in upwards of a 50 percent loss. Sensor Technology of Chatsworth, California, has produced a new cell design which embodies smaller modules producing the same amount of power as the large round cells. However, they cost less because there are fewer materials used and the cells are easier to interconnect. The price of the individual cells is somewhat reduced over that of the current round cell.

Multiple Cell

The Massachusetts Institute of Technology has developed a new cell wherein a number of 9.5-mil-thick N-type silicon wafers are fused together with a PN surface layer on opposite sides of each wafer, as shown in Fig. 7-10. They are stacked together using an aluminum foil to provide physical rigidity as well as an electrical contact. The stack of silicon cells is then sliced vertically and cut into slabs. After an etching is done to remove the saw marks, leads are attached to the silicon and a varnish is applied. These slabs become a cell with a number of PN junctions perpendicular to the front surface

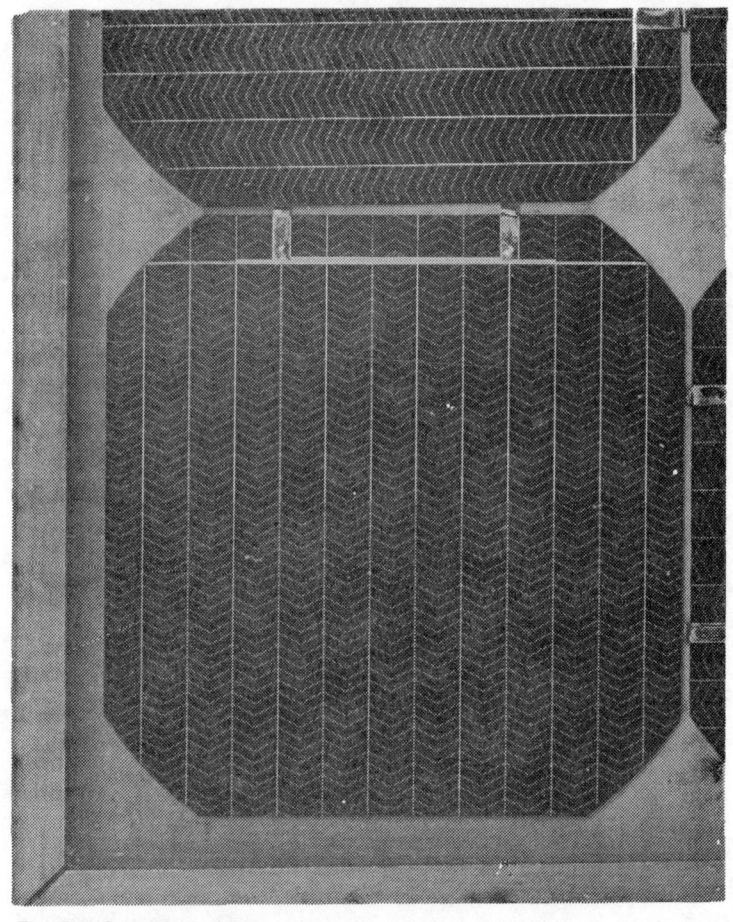

Fig. 7-9. Cell with six segments (courtesy Solarex).

at right angles to the original silicon wafer. With the contacts and the PN junctions in parallel with the radiation from the sun, the current travels through the device, eliminating the need for a contact grid which, in most cells, creates a certain amount of shading of the cell surface and also reduces the resistance in the cell itself. Tests have shown 8-percent efficiencies for this type of cell under normal radiation from the sun, but the researchers went on to find greater efficiencies by concentrating the sun. The use of concentrators will be discussed in a later chapter. Another solar energy company has announced the development of a new solar cell which converts 30 percent of the sun's radiation into electricity. This type of solar cell is made of layers of overlapping miniature cells connected in parallel. Solar-cell arrays could be reduced in size by this method, making them more practical to use on private residential roofs for electrical power generation.

Thin Films

In an attempt to eliminate much of the waste, new methods of producing thin films of silicon are being developed. These silicon-cell structures in a thin film or in continuous ribbons would do much to reduce the overall cost of the silicon cell and to reduce the waste involved in the cutting and slicing process. As improved production techniques and better materials are found to increase the solar cell efficiency and to reduce the cost, we will see the widespread use of solar cells. As a people, we have become so dependent on electrical energy that it is logical to reason that photovoltaics will be the industry of tomorrow.

New Materials

The future of this branch of solar energy is truly great. While some people are working on improving the silicon cells, others have thought that the answer lies in some new materials. Several new materials are light-sensitive semiconductors. However, the problem is to find one which will give greater efficiencies at lower cost. So as scientists have explored these materials, they have come up with several new possibilities that may help solve some of the problems. Some great technological breakthroughs may come about in the production of solar cells in the next few years. It has been demonstrated that several other materials can be employed, thus reducing the amount of silicon now going into cell production. As some of these new substances are developed into more efficient solar cells, the emphasis on using silicon in cell development may decrease.

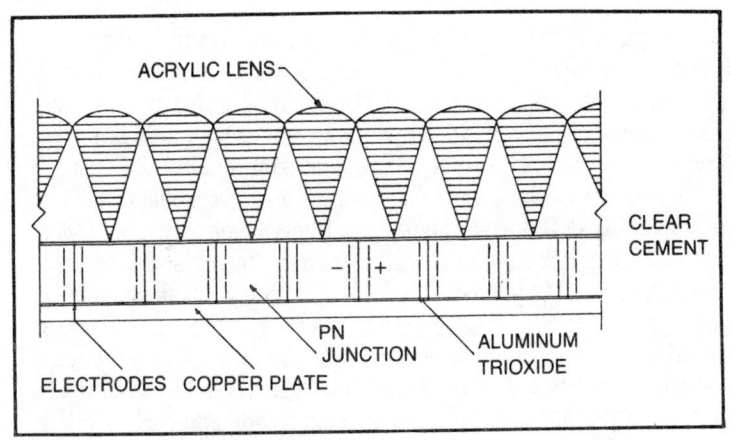

Fig. 7-10. Stacked multiple cell.

One of these new materials is *cadmium sulfide*. These cells are no longer in the experimental stages, but are currently being sold on the market. Cadimum sulfide has been cheaper to manufacture because of the simpler process by which it is produced. A thin layer of cadmium is applied to a substrate material. In comparison with the silicon cells which require about 100 microns of silicon, only two microns of cadmium are needed.

The thin-film cadmium sulfide cells have had many problems in the past, such as a breakdown of the material in a short time and lower efficiencies. But the manufacturers of these cells have made great technological advances in sealing the cells to reduce the amount of degradation and breakdown. These wafer-thin cells using crystals of cadmium sulfide are covered with a thin layer of *indium*, a very soft, semitransparent metal. In a cadmium sulfide cell, the top faces the sun. The underneath side is covered with a thin layer of silver. One wire is connected to the silver side, or positive connection; the other wire is connected to the indium layer on the top, which is negative. When a connection is made, an electrical energy can pass from one layer of the cell to the other. As the cadmium sulfide side is exposed to the sunlight, a current flows through the circuit and operates an electrical motor. So this is, in fact, a very simple and true form of a solar generator. It must be realized that the amount of energy obtained from a cell of this nature would be very minimal. However, any amount of energy could be produced by combining cells in parallel or series. Enough energy was produced to drive an electric clock from the first indium cell. This cell was only about ⅛-inch square. With improvements, this cell could yield a considerable amount of energy.

Cadmium sulfide or copper sulfide cells are fabricated in a very different way, using an electroplating process. A thin layer of cadmium sulfide is applied to a substrate of copper. Because the cadmium sulfide cells have tended to degrade under high temperatures, they must be cooled to remove the heat from behind the cells.

One of the earlier homes using solar energy was built in 1971 by the University of Delaware Institute of Energy Conservation. Called the Solar One Home, it used cadmium and copper sulfide cells to supply the electrical needs to the house. Acting as an airtight collector to provide the heating requirements for the home, hot air was passed behind the cadmium sulfide cells to carry away the excess heat from the cells. The University of Delaware's research teams produced a thin-film solar cell of cadmium sulfide and copper sulfide with about 7.8-percent efficiencies in converting the sun's energy into electricity. The results of these experiments and others could lead to a significant increase in commercial use of cadmium sulfide in the next 5 to 10 years.

Thin-film solar cells have been studied since 1961, but efficiencies have remained below 7 percent, and the cost of the cells was approximately $1 to $12 per square meter, not including processing. ERDA's overall goal for thin-film cells is to reach about 10-percent efficiencies by 1980 and to get the price down between 10 cents to 30 cents per peak watt. This goal does not seem completely unreasonable since thin-film cells are more easily mass produced and imperfections in the cells are less critical. Because of this, the thin-filmed cells appear to have lower cost potential. Three of the materials being used to develop thin cells are copper sulfide, cadmium sulfide, and amorphus silicon, all costing less than the current silicon solar cells. However, these cells have all exhibited somewhat lower efficiencies than that of silicon. Improved techniques in producing the cells could result in higher efficiencies. Many of these cells are still in the experimental stages and have produced efficiencies of only about 5 percent. In the past, degradation of cells has limited their operating cycle. With the improved stability of these systems, the potential of thin-film devices may now be actively pursued. However, the silicon cells are two to three times more efficient than the cadmium sulfide cells.

In order to stimulate the solar cell development, it is only reasonable that the government and industry directly involved in making a profit in the immediate future continue using silicon solar cells. The government and the rest of the people involved in solar energy research believe that silicon has a long-range potential of about 50 cents per watt. If this happens, cadmium sulfide and the other materials being explored will have to be produced for less than 50 cents per watt to be competitive with the price of silicon. So the cadmium sulfide and other cells being experimented with may eventually be developed into a viable product, but they will take a little longer to perfect. It is thought that by 1980 to 1990, cadmium sulfide will be a substantial contributor to the photocell field.

Experiments now being made with cadmium sulfide and copper sulfide show possibilities of increasing the energy efficiency of these cells in direct sunlight beyond the present value of 8.5 percent. But it is anticipated new experiments and continued research will eventually lead to an energy conversion efficiency of up to 14 percent for cadmium and copper sulfide cells.

A new process of spraying cadmium sulfide onto a copper or glass substrate has proven to reach 5-percent efficiencies. Efficiencies of 8 percent or better are being predicted with some confidence. Using a substrate of window glass or copper, these black-wall cells are being put on by a flotation process. A thin film is sprayed on the surface of heated glass. As the temperature of the glass cools, a very thin layer of cadmium sulfide is applied. Pilot plants are now under construction which will manufacture large numbers of these cells for commercial use. Wide ranges of compositions can be formed on thin films by allowing a vapor deposit, as well as by using such techniques as sputtering the material onto the substrate. These large areas of thin films could be utilized in place of the silicon-solar cells to produce electricity directly from the sun much more economically.

Another very different type of material being used in solar-cell production is that of modified amomorphic materials. The amomorphic materials are noncrystalline substances suitable for low-cost energy conversion. These materials have a disordered atomic structure in comparison with the crystalline materials being used in silicon solar cells. The crystalline materials used in silicon are semiconducting devices similar to those used in transistors and diodes. Currently, these crystals can be made no larger than a four-inch saucer and are very expensive to produce. A cell made of modified amomorphic materials is being explored by Energy Conservation Devices Inc. Although a low degree of efficiency is obtained from these cells, they are very inexpensive to produce. It has been predicted that these cells, given a 20-year lifetime, would possibly produce energy at 2 to 3 cents per kilowatt. And with present thin-film technology, we could produce large quantities of this type of material. The amomorphic materials are considered to be one of the newest frontiers in solid-state physics. Until now, the widespread conversion of sunlight into electricity has been limited by the cost of growing silicon-crystalline materials. The amomorphic semiconductor materials, however, do not require such careful control in the growing processes and thus become a much lower-priced product.

It is calculated that the amomorphic material could produce power at 3 cents per kilowatt. Current solar cells of the crystalline family are producing electricity at about 25 cents per kilowatt-hour. The amomorphic materials have a much lower efficiency rating of about 6 percent and possibilities of about a twenty-year life span. However, it is generally felt that greater efficiencies could be produced.

OTHER METHODS OF ELECTRICAL CONVERSION

There are many other ways of converting the sun's energy into electricity. One is the conversion of sunlight into *thermal energy* used to heat the temperature of water to steam and then to drive steam turbines that run electrical generators. Unfortunately, though, the conventional solar collector seldom gets warmer than 200° Fahrenheit. Therefore, complicated tracking and concentrating collectors are needed to run these generating plants, which must reach temperatures of 1000° Fahrenheit. As one can see, this is a much more complex and technical process than using solar cells for the generation of electricity.

Since this book is primarily dealing with the use of the sun in generating electrical power, it would not be amiss to mention another process called *thermocouples*. The thermocouple takes advantage of the sun's heat to produce a thermoelectric effect. A thermocouple is simply two wires of different metals—for instance, antimony and bismuth—joined and twisted together on the ends. When this junction is exposed to the heat source, which in this case could be concentrated sunlight, an electrical current flows from one wire through the load to the other wire. Thus, by heating the junction from the concentrated sun, and keeping the other end in a cool place, considerable amounts of electrical energy can be obtained and converted into a usable form of power.

For decades, astronomers have used instruments employing the thermocouple principle, many of which are so sensitive to light that they can measure the intensity of a candle from a distance of 100 miles. When the junction of the two dissimilar metals is heated, they produce an electric current. By focusing the light of a star on a thermocouple, astronomers were able to determine the temperature of the star by the amount of current it produced. The thermocouple technique has been employed for some years by gas-fired

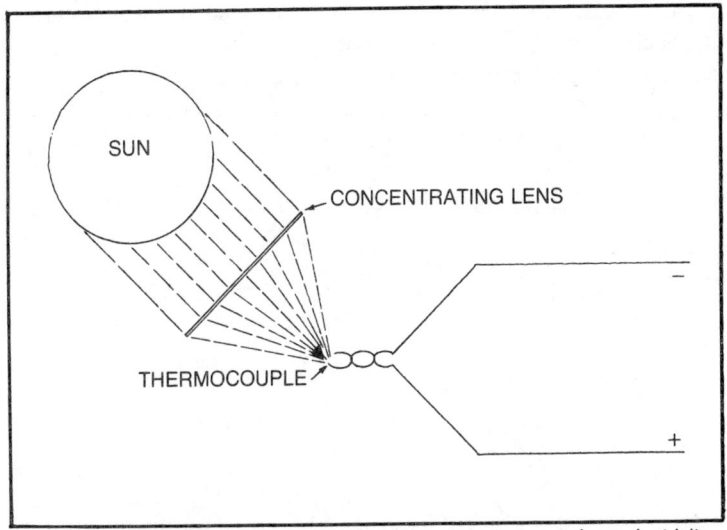

Fig. 7-11. Focusing the sun on a simple thermocouple to produce electricity.

generating systems in remote applications, but the use of concentrated sunlight on thermocouples offers some attractive possiblities. In both applications of thermocouples and photovoltaic cells, electrical current can be obtained by the sun.

A simple thermocouple project can be made by twisting a piece of copper wire and a piece of steel piano wire, as shown in Fig. 7-11. As the copper wire is hooked to a sensitive meter, and the other side of the meter to the steel wire, a current is produced which can be measured by a meter. This current can be increased by imbedding one side of the junction in a cup of ice as the other side is heated by a gas flame or by concentrating sunlight. The amount of voltage produced by such as apparatus is very low, but it may be possible to combine it with other thermocouples. If many were used, a large amount of voltage could be produced.

Much experimentation has been done to find the best combination of metals to be used in this type of system: copper-constantan, iron-constantan and platinum-platinum rhodium have all been successfully used. When several of these thermocouple units are hooked together, they are called thermopile and can generate a large amount of electrical energy. It is predicted that considerable amounts of electricity could be produced if a mile of this type of thermocouple were coiled together in a small space, and more than an acre of large mirrors were focused onto this junction. The problem with using thermocouples is the large amounts of very scarce metals that are used. Some scientists hope that thermocouples could be made with more common metals that could develop greater efficiencies. As experimentation continues, combining the solar cell and thermocouple principles could produce a better cell.

Just a few methods of production and materials used in solar cells have been discussed in this chapter. In future years, many more ways will be developed. Whatever materials or methods are used in the development of solar cells, their future looks promising.

Concentrator Cells 8

Due to the high price of today's cells, many attempts have been made to reduce their cost. The goal is to achieve lower costs for photovoltaic power conversion. One approach has been to concentrate the sun's light so that fewer cells are needed. To do this, the sun is concentrated many times and focused on one cell. So, a special type of cell has been developed to work efficiently under the concentration of many suns, as shown in Figs. 8-1 and 8-2. The use of concentrators to focus the solar energy on the cell with more intensity holds great promise for lowering the cost of solar cells in the immediate future.

Several companies have produced cells rated for 20 suns. One company has developed a cell that will operate under the exposure of 100 suns. By the use of concentrating solar-photovoltaic collectors, overall costs have been reduced by one-half on larger installations.

At present, the concentrating photovoltaic array offers significant advantages over the nonconcentrating arrays because it reduces the required number of presently very expensive solar cells to produce a given amount of electricity. The main emphasis in the concentrating device is the reduction of both the cell area and of the materials involved in the production of photovoltaic cells (see Fig. 8-3). Thus, a trade is made between the cost of the cells and the cost of the concentrating system. However, when the price of solar cells is reduced to 50 cents per peak watt, the use of concentrators will actually be more expensive than the use of additional cells. When this occurs, the concentrating solar cells will basically be phased out.

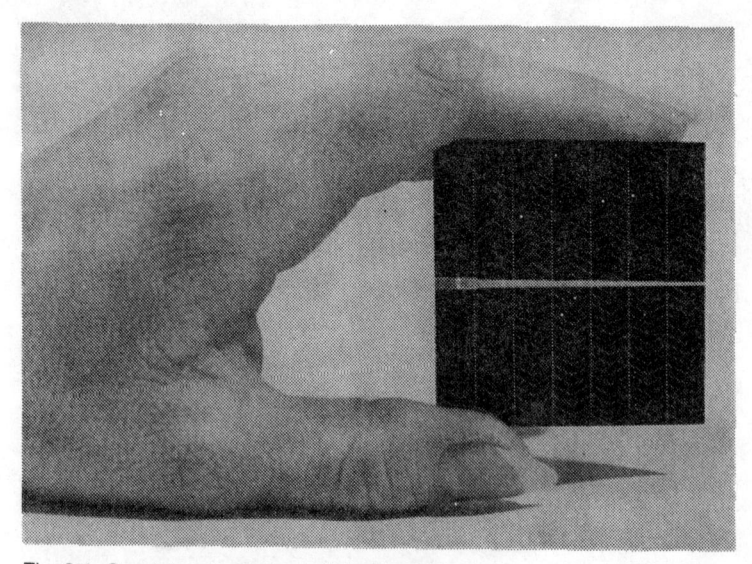

Fig. 8-1. Special concentrator cell (courtesy Solarex).

However, it will be some time before the price of photovoltaic materials can be substantially reduced to eliminate the need of concentrators. Until then, the concentrating device may have a significant impact upon the photovoltaic market.

The reduction of the total solar-array cost is the main reason for using concentrating arrays. The use of these arrays with solar cells

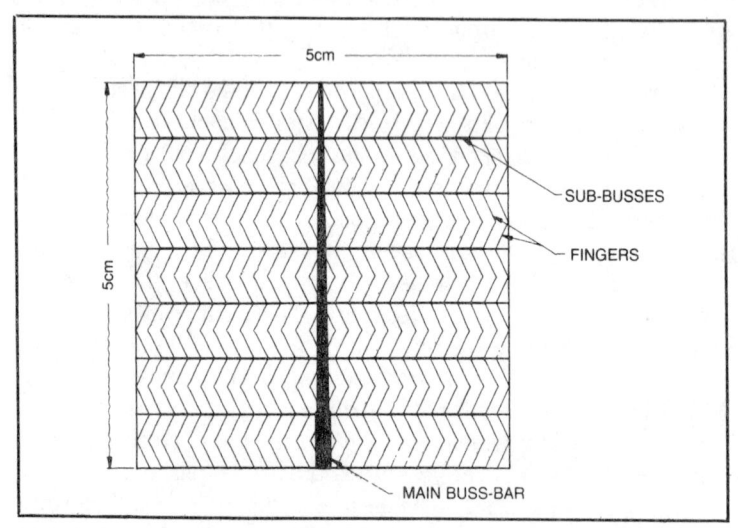

Fig. 8-2. Physical dimensions of a concentrator cell (courtesy Solarex).

Fig. 8-3. Low concentration arrays (courtesy Solarex).

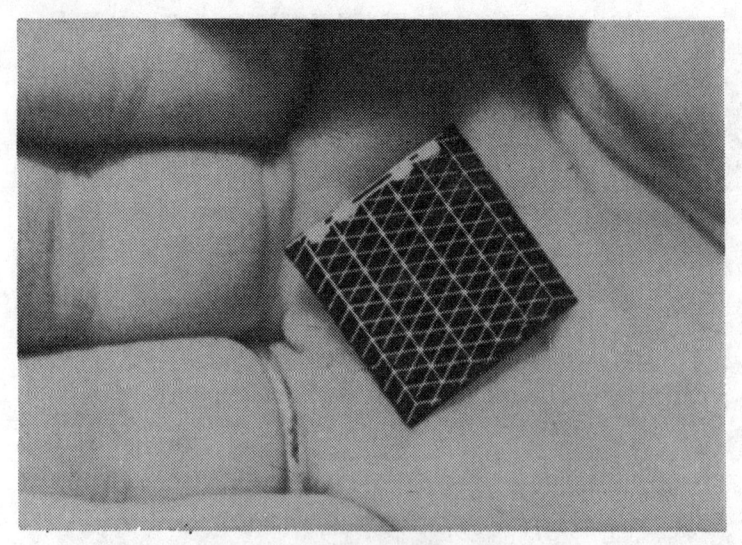

Fig. 8-4. Mounted in a sun-concentrating system, this concentrator cell generates 2 watts from only 4 square centimeters of area.

shifts the emphasis with respect to technology from cell cost to cell performance. High conversion efficiencies will therefore have top priority in the concentration-cell research and development.

THE CHARACTERISTICS OF CONCENTRATORS

Concentrators have been used for years to focus the sun's rays, producing very high temperatures. Many a boy has used a magnifying glass to focus sunlight on a piece of newspaper, setting it on fire. This, then, is an example of concentrated heat. There is more to the sun than mere heat, as is evident through observation of the solar cell. The solar cell obviously uses something more than heat, for as the temperature increases, the efficiency of the cell decreases.

The concentrator cell is exposed to sunlight as any solar cell, but instead of each cell receiving power from one sun, the concentrator cell focuses a larger area of sunlight on one point. A concentrator cell placed on this focal point will produce as much electricity as a group of standard solar cells covering the entire area of the concentrating lens. With the high cost of solar cells, it is possible to produce electrical power with less material by using a concentrator, particularly when the concentrator lens costs less than the equal number of solar cells.

A concentrator cell works much like that of a regular solar cell. It, too, converts the sun's energy into electricity. The cell is still made from the same basic materials, and must be grown, cut and doped like the regular cells. But the concentrator cell differs in many ways. The concentrator cell is made with a different grid configuration as well as a more highly-controlled doping process (see Figs. 8-4 and 8-5). Because the cell can produce a higher electrical output, the concentrating cell requires a much finer, closer grid pattern on the cell. In the production of solar cells, many different grid patterns have been tried to find the optimum pattern that will harness the most current. And with the advent of the concentrator cell, it was found that different grid patterns are even more critical. The reason for the difference in the grid design is the increased power buildup. The closer grids are needed to carry away the additional increase in power. More sophisticated materials are required because when the contacts of these grid patterns come too close together, the potential of the tinning bleeding over is possible. Accidents and excess material buildups become more likely.

Materials

Three distinct materials are being actively considered for use in photovoltaic systems. These are silicon, cadmium sulfide, and gallium arsenide. Cadmium sulfide and silicon will probably continue to be used in flat-plate systems with no concentration. However, owing to the relative scarcity of gallium arsenide cells, their most wide-

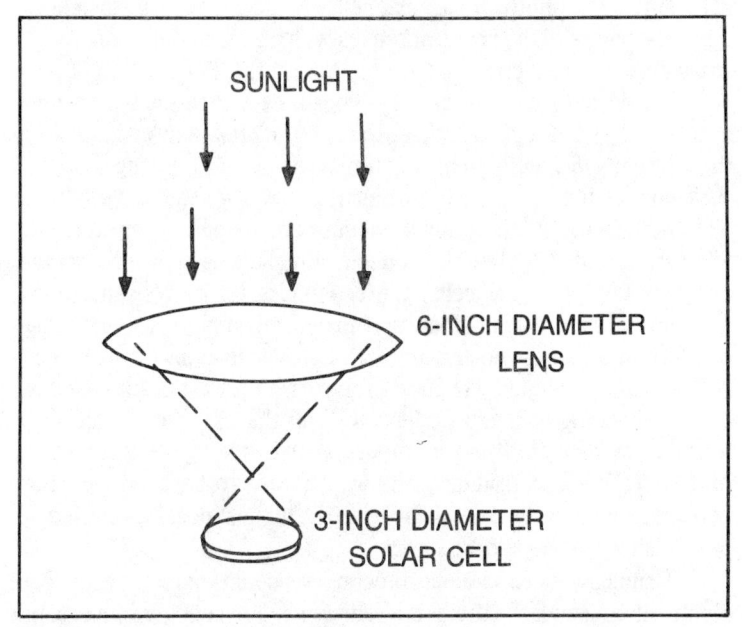

Fig. 8-5. Principle of focusing to concentrate light on a solar cell.

spread application will most likely be in central electric generating stations where concentrating systems can be utilized best.

Mounting

In addition to the variance in materials and grid designs, the cells must also be individually mounted to suit the application. In most cases, the cells are attached directly to a backup plate. This backup plate is then connected to a pipe to carry away the excess heat. Several mounting techniques have been developed. One company has developed a cell with a hole in the center, allowing for a more positive method of mounting.

TRACKING DEVICES

For a photovoltaic concentrating device to work properly, most must follow the sun across the sky so that the lens is always focused on the cell. Due to the nature of the earth's rotation, the sun spends a good deal of time in positions where its rays do not shine directly on the surface of a fixed flat-plate solar array. If the array's efficiency remains constant, up to 50 percent more energy per year may be obtained by a given array when tracking is used. This may vary slightly, depending on the climate.

With a properly designed system, the gain could be even more substantial, and photovoltaic solar cells should increase in efficiency; thus, the output of the array increases with greater solar intensity. A small amount of electrical current is lost in the PN junction of any cell. This loss does not increase proportionally with the increase in current. By using optical concentration to increase solar intensity, the average efficiency can be further increased in a tracking system. Efficiency is not the only determining factor in a solar cell system. The real factor is the capitol investment involved per average kilowatt output. So when the concentrating lens is cheaper per unit aperture than the solar cells, the economics favors concentration. Concentration will almost always prove advantageous when the converting device is a solar cell. Many new applications for solar cells will be made possible by reducing the current price through the use of concentrating or parabolic lenses. With the use of concentrating lenses, one can afford to pay more per unit due to the increase in efficiency. When considering the type of cell to be used with the concentrator, the cell that will provide the highest possible efficiency per dollar spent should be chosen.

If sunlight were monochromatic, we could expect to convert electrical energy from the sun in efficiencies exceeding 50 percent (see Fig. 8-6). But because sunlight has a broad color spectrum,

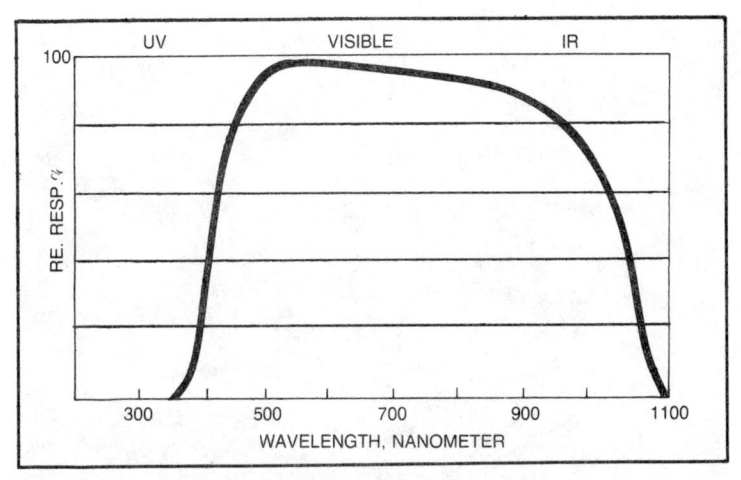

Fig. 8-6. Spectral response of a solar cell (courtesy Solarex).

much of the power of the sun is unused by the solar cell, a fact which reduces its efficiency. Because of the sun's spectrum, there are actually two different materials which have the highest efficiency for collecting the sun's energies: silicon and gallium arsenide. At room temperatures, and one-sun ratio, if silicon were considered to have 100-percent efficiency, gallium arsenide would have 110-percent efficiency. In other words, the gallium-arsenide cell is about 10 percent more powerful with everything else being equal, which makes the gallium-arsenide cell more practical for most applications. This is particularly true where thermal heat, as well as electrical energy, is desired in a system.

All solar cell devices are greatly improved by their ability to track the sun. More electricity could possibly be produced with tracking systems (see Fig. 8-7). Because of this increase, there has been much attention given to simple tracking devices. One such tracking device requires a fluid such as freon, or some other easily heated liquid. This fluid expands and goes from one chamber to another, producing a weight balance, moving the device as it follows the sun. Because of the movement of the sun across the sky, a shadow is cast upon one cylinder. The other cylinder would begin to get hotter and would push more liquid to the shaded side, thus moving the tracking device until it was more in line with the sun.

Another type of tracking device generally employed uses some type of motor in conjunction with a V-shaped photovoltaic cell arrangement (see Fig. 8-8). A shading device going up through the middle casts a shadow on one side of a photo cell, creating an electrical imbalance. This triggers a switch to drive a motor, which

Fig. 8-7. Freon tracking device.

will then track the sun until a balanced state is achieved between the two photovoltaic cells.

The other system is that of a timer which starts the motor every few minutes and moves the module tracking the sun. However, this type of system must be adjusted periodically throughout the year as the time of day that the sun would be in a certain place would not always be correlated.

It may be a simple process to use a tracking device on a small solar array, but tracking becomes more complicated as the size of the system increases. Some of the proposed large photovoltaic systems have simply used many small arrays with individual trackers on each array. One can see that this would be rather expensive and increase the maintenance involved in the mechanical nature of the tracking device.

Several simple tracking systems are designed for use with concentrating cells. One of these is a disc made of a lightweight material, such as styrofoam, floating on a shallow water pond. Some have proposed using large arrays mounted on this Styrofoam base (see Fig. 8-9). By using a sensor device which triggers the motors, the entire floating pad would track the sun with the motors turning on and off to position the pad in the sun's path. This floating disc can be moved with practically no resistance to the proper angle for the collectors to receive the maximum amount of sunlight. In such cases where the solar pond has been used, water is simply pumped from the pond below the Styrofoam through pipes attached to the solar-cell array and returned into the solar pond. This circulation of water provides not only a way of getting rid of the excess heat but is also an excellent storage medium for the thermal energy extracted from the solar cells in the form of hot water. This would make the pond underneath into an insulated thermal heat-storage unit. These rotating ponds could be relatively small in nature, or could be very mammoth in size. The use of this type of tracking device could greatly reduce the cost of a photovoltaic system.

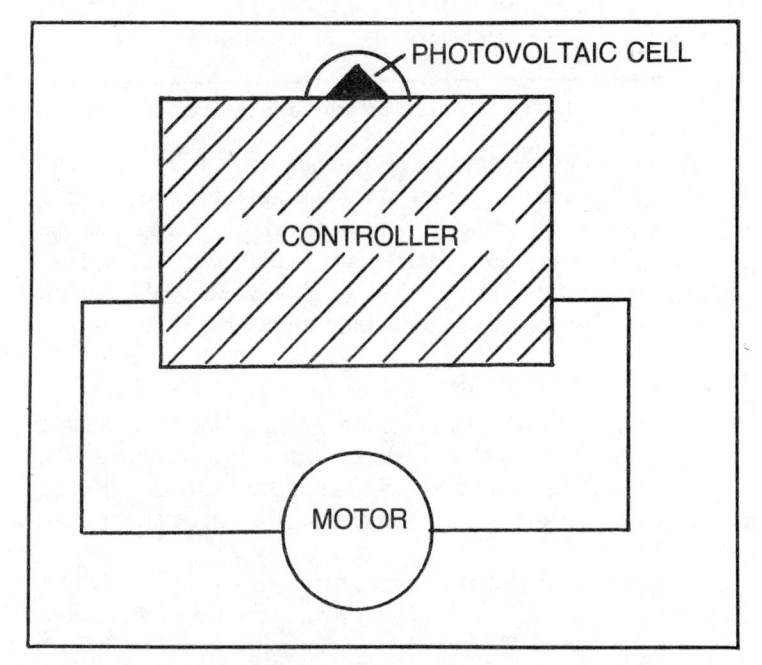

Fig. 8-8. V-shaped photovoltaic cell tracking device.

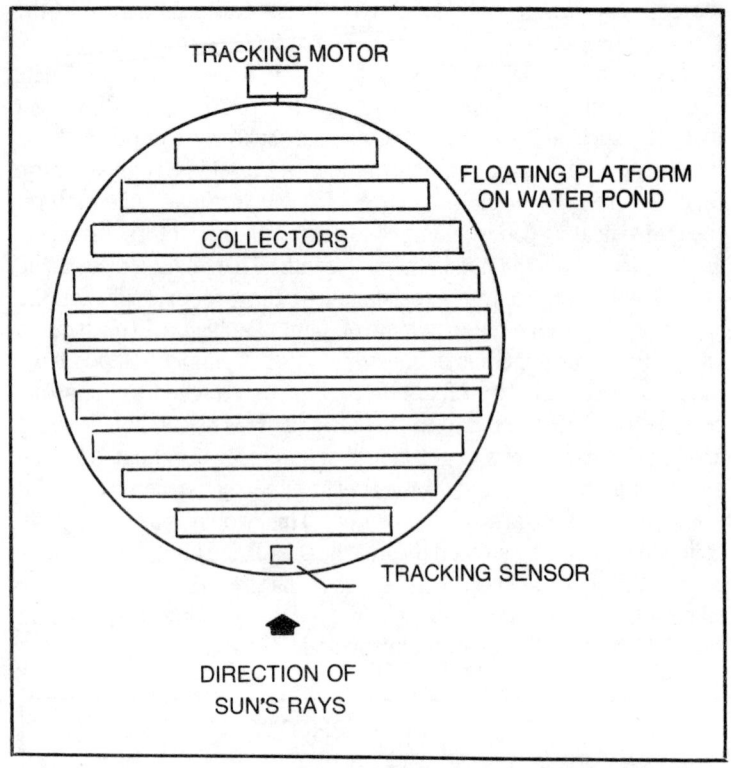

Fig. 8-9. Floating Styrofoam base tracker (courtesy Sensor Technology).

Whether concentrators are used or not, the viability of tracking the sun greatly increases the amount of voltage that can be expected from a given module through a 24-hour period. The need for simple and uncomplicated tracking devices is still very apparent. As these tracking devices are produced, they will greatly reduce the number of modules or cells required to fill the load needs.

TYPES OF CONCENTRATORS

Concentrators fall into two main groups. One type actually reflects the sun from its surface. An example of this is a mirror. The other type magnifies the sun's rays by using lenses. Both reflectors and lenses have been used in concentrating the sun on photovoltaic devices.

The use of concentrators dates back many years. Early men probably used the reflector, employing a shiny piece of metal or other similar objects. Today, we have developed mirrors with very high reflectance. Many small mirrors can be focused on one point, producing light and energy equal to many suns. Each of these mirrors must be focused and mounted separately. There is, however, a simpler method of reflecting the sun's rays: the parabolic trough or disc, in which the entire unit is designed to focus at a given point or along a fine line. The parabolic trough is made of a highly reflective material that reflects the sun into a narrow strip (see Fig. 8-10). In this instance, the cell is generally mounted on some type of a pipe that extends from one end of the trough to the other. Water is circulated through this pipe to reduce heat buildup in the cell. If this heat were not withdrawn, it would tend to lessen the amount of power which could be obtained from the cell. The light is reflected along the entire length of this pipe to which the cells are attached. A reflective disc is currently being used in large arrays being developed and designed by Motorola for use at the airport in Phoenix, Arizona. This type of concentrator focuses the sunlight onto a mirror in front. It is then reflected back down to the cell, which is mounted in the center of the disc, as shown in Fig. 8-11.

Another way to produce more power with fewer cells is to focus the sun's rays from a large area to the small area of the individual cell. It was not until man had developed sophisticated lens-grinding equipment that magnifying the sun's energy in this manner was made possible. Over the years, many methods had been employed to do this. Today, however, new and simpler lenses have been developed. One such lens is the plastic *fresnel lens*, which focuses the amount of sunlight falling on, for example, a square foot of area to a narrow

Fig. 8-10. Parabolic trough (courtesy Solarex).

pinpoint, which can be focused on the cell itself. With the development of the fresnel lens, cheap solar concentrators were made possible.

The fresnel lens is a thin plastic sheet that has been cast or pressed into a shape much like that of a phonograph record. In fact, it looks like a record because of the similar grooves, except it is clear and has each of its grooves cut on a different angle so each groove focuses the sun on the same focal point. Either a circular spot or a long thin line can be achieved, depending on the lens. Sandia Labs in Albuquerque, New Mexico, is now conducting tests using fresnel lenses to focus the sun's energy on solar cells. They developed a rack frame which uses lenses and photovoltaic cells to track the sun. Their testing facilities are available for private manufacturers' use.

One of the most attractive possibilities for improving solar cells is to use several in conjunction; each cell could pick up different parts of the sun's spectrum. This procedure would improve efficiency about 30 percent over standard cells. The solar spectrum could be split by the use of optical lenses, or cells could be stacked one on top of another, letting the cells themselves act as a filtering system. The use of multiple solar cells with concentration could achieve efficiencies high enough to bring about more extensive usage of the cells. By the use of good PN junctions and hetero junctions, the entire spectrum of the sunlight's energy could be more readily utilized.

In developing and testing, a multiple cell was made by sandwiching cells together so the PN junctions were perpendicular to the sun's rays. The research group at the Massachusetts Institute of Technology went one step further in designing a microscopic piano cylinder lens, which was especially formulated to have a focal point in the center of each cell segment. The lens was bonded to the cell with a transparent cement and was then oven cured. When these cells were tested with lenses attached, efficiencies of up to 21 percent were produced. These cells have been developed for mounting on heat exchangers to extract the excess thermal energy. These units are made of 24 separate subunit cells connects 25 cell segments. Each subunit yields approximately 12 to 15 volts. It appears much like a small butcher block table.

CONCENTRATOR BY-PRODUCTS

Reflecting or concentrating sunlight can achieve higher overall efficiencies. However, the excess heat buildup causes the efficiency of individual cells to drop drastically. Therefore, this excess heat must be carried away from the cells in order to obtain maximum

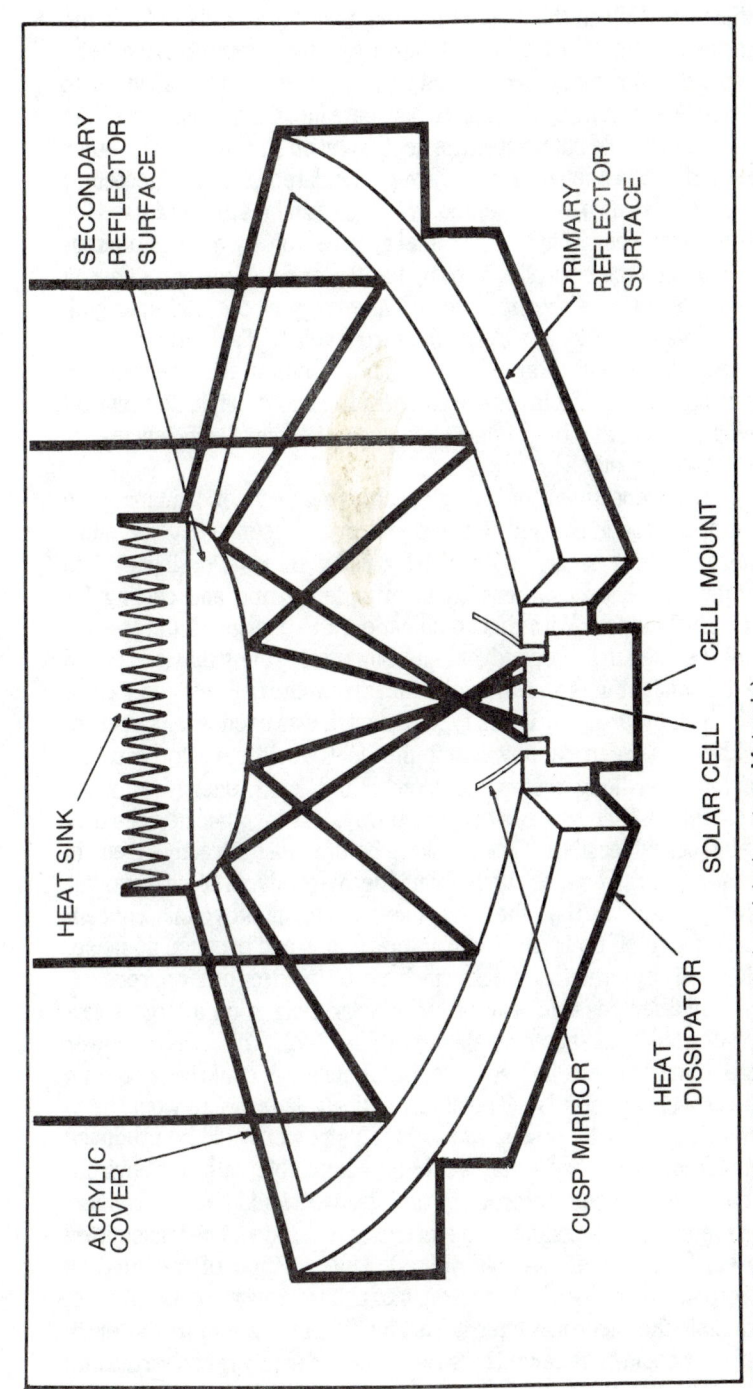

Fig. 8-11. Cross section of the Meinel optical module (courtesy Motorola).

efficiency. Unless this heat is carried away, it actually defeats the purpose of the concentrator. Therefore, two methods have been employed to remove this excess heat. One uses fins to allow air to cool the cells, causing the heat to dissipate into the atmosphere. This method is simple but wastes the heat given off by the cell. The other method is to circulate water through insulated pipes connected to each cell. By using this method, this heat can be stored or used for other purposes. This thermal energy given off by the liquid circulated behind the cells allows for up to 50 percent of the sun's energy to be converted either into electrical energy or thermal energy.

Obtaining heat as a by-product from solar cells has been a very attractive consideration. Efficiencies of 12 percent to 20 percent are typically obtained from most concentrating photovoltaic devices. By using the heat as a by-product, however, the overall efficiencies are increased greatly.

The Department of Energy is studying ways of combining the use of photovoltaics and thermal energy. Concentrating the sun's energy onto a single solar cell by using mirrors or lenses and removing the excess heat could provide heating and cooling for individual homes. With this combination, the energies from the sun could be used to greater advantage. Since everything that is exposed to the sunlight gets warm, excess heat from the cells may as well be used as a by-product. If this type of device was used on a housetop, neither the electrical power nor the heat would have to be transported over long distances. Converting solar energy into the greatest possible usage is one of the ways photovoltaic devices may be financially feasible. This seems to be the most attractive trend in the long run. The nice thing about the by-products of photovoltaic concentrators is that they are useful. The photovoltaic concentrators give off no harmful exhaust or fumes and produce no noise. Their only by-product is heat, which can be used for other purposes.

Another possible way of using solar cells is on a large solar-power station in space, as shown in Fig. 8-12. This type of power station would not be limited to daylight hours, but could be producing power 24 hours per day. The object of this solar satellite would be to provide large scale power on earth. This power could be produced on a competitive basis that would be compatible with our environmental restrictions. In order to do this, we would have to convert vast amounts of sunshine into electricity which could be transmitted to earth, retrieved and distributed. One method of transmitting electricity that is being discussed is to use microwave generators to transmit the electrical energy received from the sun to the earth where it could be received by an antenna and safely reconverted into

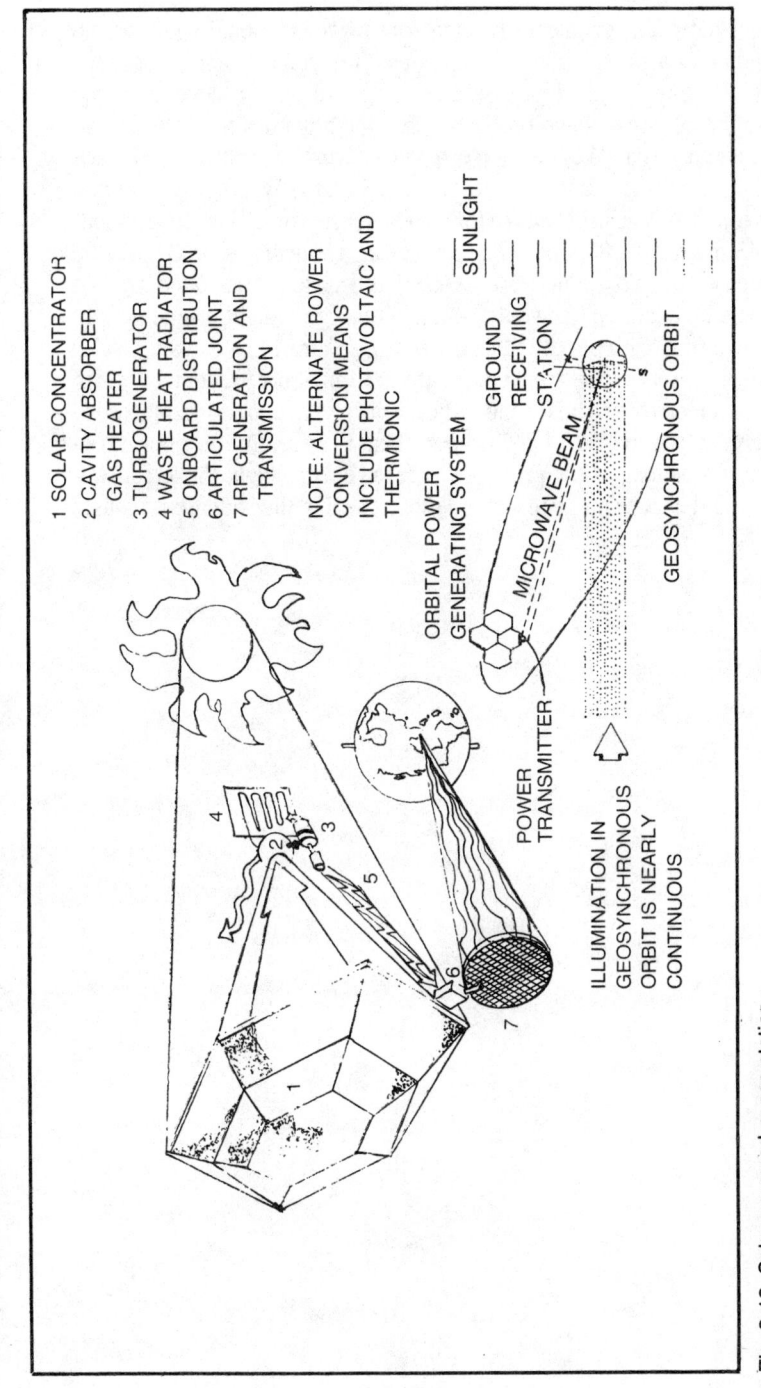

Fig. 8-12. Solar-powered space station.

electricity. The primary advantage of a power station in space would be that for a given number of solar cells, we would be able to produce 6 to 10 times as much power. In other words, to produce the same amount of power here on earth, we would need 6 to 10 times the collecting area. We also have a very favorable situation in space which would eliminate the problem of adverse weather conditions. The collector area proposed for a solar space satellite would be about 2.48 miles by 6.83 miles (4 km by 11 km) in order to produce 5000 megawatts. To reduce the area of the collectors, concentrators would be used, which would be made of a lightweight plastic film. In comparison with photovoltaic generating plants here on earth, solar satellites seem to be very competitive. In an effort to reduce costs of solar electricity, these and other efforts have been explored to increase the individual cell output. This is a step in the right direction, for as long as the price of the cells remains high, these concentrating devices can be produced more cheaply than additional cells.

The Future of Photovoltaics

9

The future of photovoltaics looks very exciting, but as with most new products, there are many things that may hold back solar electrical-power systems. First, solar systems are new, and most people are unfamiliar with them. There is often a reluctance to make a change from old to new. Second, it may take years of research to get the systems developed to the point that the public will accept them. Third, a significant reduction in the price of photovoltaic systems will be required to bring them to a competitive level.

Even though there are solid reasons for switching to solar power, it will take time to convince the general public. Decreasing cost and stimulating ideas would be highly important factors for increasing production. Time will be required to reduce costs to a publicly acceptable competitive level. Efforts by the solar-energy industry and government to stimulate the acceptance of the solar electric systems could also help, but just how fast can these price reductions be made? Acceptance and demand play a great role in answering this question. Yet, solar-cell-array prices could drop to $2 per peak watt within the next three years if the demands were high enough. The concentrator-array concept could be used to reduce this short-term goal even more quickly.

With these price reductions, the next few years will open up a whole new class of solar-powered appliances and products. We will begin to see new remote-powered applications in larger sizes, many agricultural uses and commercial appliances. There will be tremendous impact on the vacation and recreation-home industry. And in many cases, we will see solar cells being used to power highway signs, safety signals and emergency telephone stations.

In 1970 the cost of solar cells was about $200 to $300 per watt. By 1975 the price was down to $20 to $30 per watt, and by 1978, only three years later, the cells had been reduced to about $12 to $14 a watt. In many cases, the price reduction through large-scale buying has brought the price to $3 to $12 per peak watt. By the mid-1980's, the price will probably be down to about 50 cents per watt. In the coming years, the price of solar-electric systems is expected to be reduced even further. A study made by Spectrolab in California showed that in 1986, the Energy Research and Development Administration's goal to reduce solar-cell arrays to 50 cents per peak watt would bring solar-power-generation costs close enough to conventional power prices to justify large-scale electric usage. Spectrolab's study also showed that one of the first systems to be competitive will be the intermediate-sized onsite power plant, mainly because the limitations of high fuel costs, energy restrictions, local ordinances and power distribution will create a competitive situation that will cause people to look seriously at solar power.

The production of solar cells has doubled over the past year, and the Federal Government is basing its program on more than a twofold increase each year for the next 10 to 12 years. If the industry increases at this rate, one of the major problems will become the lack of experienced laborers to install and maintain the systems.

In an effort to convince the public of the feasibility of photovoltaic power, immediate and well-planned demonstration programs are needed. ERDA is developing a program to help reduce the cost of solar cells and to increase public awareness. This government agency has funded several large photovoltaic-generating facilities in hopes to stimulate the market enough to make it possible for companies involved in the production of solar cells to reduce their prices. Their goal is "to identify potential attractive applications for solar-electric systems related to current and future needs, to assess economic factors and market impact on each application, to determine the operating characteristics of applications and significant potential, and to confirm by testing and demonstration that photovoltaic conversion systems will work for these particular applications."

As one studies the exciting potential of solar-cell power production, the question may be asked, "Why is this industry not more widespread and more highly publicized?" The answer has to do with the political situation of the federal government in relation to the large corporations of the United States. In this time of energy crisis, the solar-cell development is not receiving the help from the federal government that it should in comparison to some of the other programs, such as nuclear energy. For example, in 1975 nuclear energy received $1.5 billion from the federal government, while photovoltaic research and development was given about $8 million, a comparatively small amount. In 1976 ERDA sprouted from the old Atomic Energy Commission, and until recently, most of its members were in the nuclear field.

Because the solar-cell industry is relatively new, many of its manufacturers are small businesses, and sometimes it takes the government a while to recognize them. Some of the large companies in the Unites States are waiting for the price of solar cells to come down before they want to get involved. They are waiting for much of the research and development work to be completed before they want to commit themselves. And because of the interests, many of these large companies would not benefit by the widespread use of solar cells. Their apparent lack of interest could become a serious problem for the growth of the solar-cell industry. However, a few large companies are now getting involved in this field.

GOVERNMENT INVOLVEMENT

One of the primary goals of ERDA is to develop a program to show practical solar applications for widespread commercial acceptability. Being investigated are four major areas. First, the direct thermal applications of solar energy, including solar heating and cooling, for buildings and industrial processes and applications; second, solar-electric applications, including systems based on solar-thermal electrics, photovoltaics, wind, ocean and energy conservation; third, the fuels from biomass, including utilization of land and ocean production of plant biomass as well as the collection and utilization of organic components of agricultural and forestry wastes; fourth, the technological support, including solar-research assessment and information dissemination. The impact of solar energy technologies has been estimated to be about 1 percent by 1985 and will be up to 8 percent by the year 2000, and it is predicted that 25 percent will be provided by solar by the year 2020.

The development of ERDA's National Photovoltaic Conversion Program is part of the government's overall solar development. The cost of solar cells can foreseeably go down to 1/50 of their current cost. The low-cost silicon solar-array project, called LSSA, is a division of ERDA and is part of the Solar-Photovoltaic-Development Program, which was organized in 1975. This program was directed toward satisfying the national energy needs with commercially viable solar-photovoltaic power systems. The goal of the Department of

Energy's (DOE) Program in photovoltaic conversion is to solve some of the technical and economic problems retarding the widespread use of solar cells.

LSSA is essentially an organization divided into two major task areas. The first is to analyze and solve the problems of realizing the goal of 50 cents per watt photovoltaics, compared to today's price of about $15 per watt. In order to accomplish this, some of the following areas will have to be explored: first, the reduction of the basic silicon material cost from $65 per kilogram to less than $10 per kilogram; second, the effective conversion of these materials into large-area silicon sheets suitable for the production of high-efficiency solar cells; third, the low-cost technology to convert the sheets into arrays; and fourth, the development of long-life encapsulation methods to protect solar cells from this earthly environment for periods of more than 20 years.

The other objective is to develop a commercial readiness to produce arrays of fewer than 500 kilowatts of peak solar-power by the year 1986 and to expand the annual production of the silicon-solar-cell modules from the current level of about 100 peak kilowatts per year to that of 500. ERDA is now seeking a viable solar program for the development of economical solar cells to provide a sufficient portion of the energy needs of this nation by the year 2000. Solar-power arrays costing less than $300 per peak kilowatt are expected by the year 2000. The goal to be reached by the year 2000 will perhaps use photovoltaic devices other than what present technology may afford.

The price of solar cells has been reduced dramatically in recent years. But DOE's goal is to bring the price down still further so that by mid-1980's solar cells might be making a substantial contribution in meeting the energy needs of the nation.

DOE has identified two types of applications for solar-electric plants. One would be the large power plant in the utility network. The second would be generating facilities in areas of remote applications. For many remote applications of less than 100 watts, solar-electrical plants are already commercially competitive. Many of these systems are presently being marketed for $30 to $70 per peak watt, including the array, solar storage batteries, regulators, and power conditioners. The solar cells are especially suitable for small applications because they may be built into many small units. Up to a certain point, photovoltaic power plants become more economical as they become larger; however, the optimum size still seems to be smaller than the usual size of a nuclear or fossil-fuel power plant. But

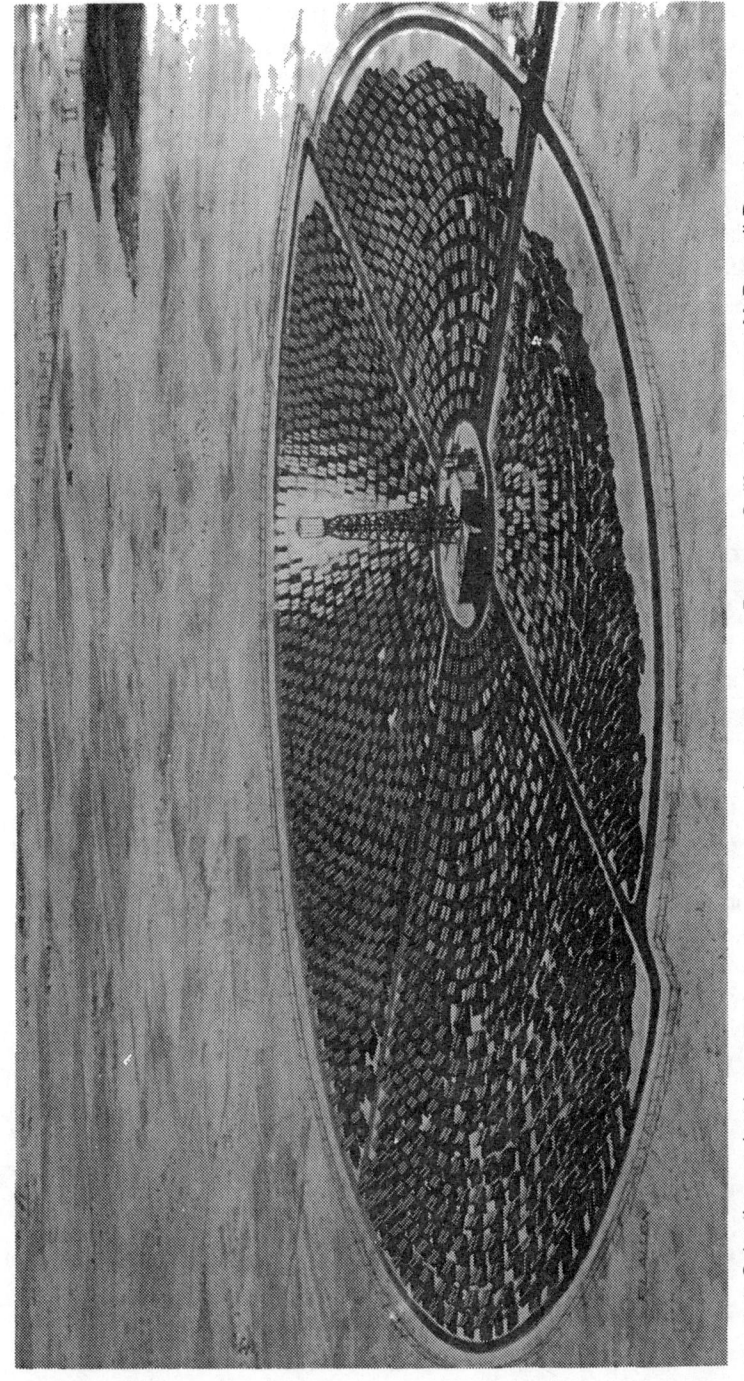

Fig. 9-1. Solar thermoelectric power plant using mirrors, under construction at Barstow, California (courtesy McDonnell Douglas).

large-scale applications will require lower prices, and DOE's efforts are aimed at bringing the price down to a point where these large applications will become possible. When this happens, the electricity from photovoltaic cells will begin to contribute significantly to meet the nation's energy needs, especially since the cost of other forms of generating electricity will probably be on the increase.

There is little doubt that within the next 10 to 15 years, the solar-cell production of electricity could become competitive with other methods of electric generation. But the cost reduction in the near future will depend greatly upon the government, political policy and public acceptance.

To get some idea of the government's involvement in the area of solar energy, let's look at the President's fiscal budget for ERDA in 1977. That year $116 million was alloted for ERDA, compared to $86 million in the year 1976, showing an increase of about 35%. The breakdown for the year 1977 into four program units shows 32 percent of this amount going to direct thermal applications, about 62 percent to solar-electric applications, and about 3 percent to the areas of biomass and technological support. So you see from this that great interest is being given to solar-electric applications. The object of the Solar-Electric-Application program is to develop and demonostrate technologies for collection and conversion of solar energy into electrical energy to make possible annual energy contributions before 1985. The first of these solar-electric applications is wind-energy conversion; the second is solar-photovoltaic conversion, involving direct conversion of sunlight into electricity through the use of the photovoltaic process; the third is solar thermal-electric conversion; and the fourth is ocean thermal-energy conversion.

To accomplish these goals, one of the program milestones is to develop a 1-to-10 megawatt electric wind-generating system by 1979 to 1982. The second is the cost reduction of photovoltaic elements by a factor of 40 by 1985. The third is to have a 100-megawatt electric solar-thermal demonstration plant operating by the mid-1980's (see Fig. 9-1), and fourth, to have a 25-megawatt ocean-thermal pilot plant in operation by the mid-1980's.

All of these are ways of producing electricity from the sun. The sun provides the source of fuel which is constantly being renewed. The economic competitiveness for these applications makes solar-electric technologies a high priority in ERDA's overall development and effort. The photovoltaic program is well on its way because photovoltaics have already been proven viable in many current applications.

In 1975 production of solar cells for terrestrial purposes exceeded the production of space-quality cells for the first time. Many of these systems are now in successful service. Several of the energy needs of this nation could be filled today with some type of solar-electric system. But the challenge today is to identify the appropriate uses for photovoltaic power and to develop the market for the new products powered by solar cells.

LOOKING AHEAD

What will the future hold for these fascinating little photovoltaic wafers? Today's costs are high, but manufacturers are predicting that in the near future, with improvements in manufacturing and inflation of other energy sources, the cells will become competitive. When this takes place, solar cells will be used for powering homes,

Fig. 9-2. Photovoltaic module (courtesy Solarex).

recreational vehicles, cabins, communication systems, cars, and public transportation vehicles (see Fig. 9-2). The sky will be the limit for the applications of this little giant.

Engineers are expected to consider plans for solar-cell power stations in the early 1980's. These stations would allow us to reduce our dependence on fossil fuels. We may begin to see power stations in the 1-to-10 megawatt range by the late 1980's. We will see communication systems using photocells, and see large industrial plants capable of generating more than 10 megawatts. The solar cell industry will be producing 5000 megawatts of solar cells by the 1990's. There has been some talk of power networks in the 100-megawatt range. Production of solar arrays could reach 10,000 megawatts by 1995 and 50,000 megawatts by the year 2000. It is estimated that by the turn of the century, this nation's total electric-power needs will be 16,000 million kilowatts of power a day. Solar electricity, it is predicted, could supply some 3 percent of the total by that time.

Several studies have been made to assess the potential role of photovoltaics in future generating plants. In a study by General Electric, several United States utility systems were compared to the photovoltaic generating system as a method to evaluate the viability of the photovoltaic system. In these studies, estimates were made of the value of photovoltaics to the utility companies. At present, photovoltaic devices are both technologically and economically feasible for reliable large-scale applications. Many promising new approaches for the use of photovoltaics in the future are becoming apparent.

There is an effort on the part of the government to provide program direction and to set goals for photovoltaic research and development. To accomplish these objectives, certain priorities and key components to the large-scale central applications need to be established. Because modular components can be used in photovoltaic power plants, these goals must apply to a wide range of power requirements.

The general public is now beginning to see photovoltaic cells in use in small electrical applications in remote areas for buoys and radio repeating stations (see Fig. 9-3). A large portion of the public, however, is completely unaware that such devices as photovoltaic cells even exist. In the near future, many changes will occur in this field. Through the government programs and those of private industry, the public will gradually become aware of photovoltaics and their tremendous applications in providing electrical energy in the future.

Fig. 9-3. Solar radio repeater station (courtesy Solarex).

SUMMARY

Today, it seems that in the solar field, much attention is being placed on the hot water and active-type collecting systems. Driving along city streets, collectors are frequently observed on roofs of houses in various parts of the country. In the future, though, it is relatively certain that many of these collectors will be for electrical generating purposes. With these collectors, it is possible to produce enough electrical power from the average roof of a home to completely power that home with electricity to spare (see Fig. 9-4). Electricity during the day can be stored in batteries for nighttime and inclement weather use. It is possible that they could be designed into a home with less modification than is currently required for some solar hot water heating systems. In the case where solar arrays would be attached to the roofs of homes, a decentralization of electrical power would be obtained, thus eliminating the need for long electrical transmission lines across the country.

Many people who are very excited about the possibilities of using photovoltaics for large-scale generating plants. Located in the sunny desert region of the southwestern part of the United States, these plants could be connected to the conventional utility grid for transporting electricity via the existing system of power lines. Either way, use of photovoltaic facilities would greatly reduce the amount of fossil fuels needed to generate electrical power and might go a long way toward supplying the increased power requirements of this country.

Since the cost of electrical power from other sources is on the increase, it shouldn't be long before solar-electric power becomes competitive. As solar cells come down in price, their popularity increases. When this happens, many people will be switching to solar power for its many advantages. As previously mentioned, photovoltaic cells convert visible light directly from the sun into electrical energy. This power source is already being used to operate many electrical appliances and devices.

Photovoltaic conversion from an inexhaustible source offers potentials of providing significant quantities of electrical energy from the sun. This clean and silent process involves no heat or moving parts to convert solar radiation into electricity. The modular design used for photovoltaic components permits a wide range of applications, sizes, and types of collectors, using the same technology and the same basic approach. Enormous benefits to mankind could be provided by solar-photovoltaic electrical generation if it becomes practical and economical on a widespread basis.

Photovoltaic systems could reduce significantly our dependence on fossil fuels while minimizing adverse economic and environmental side effects. Although the basic photovoltaic conversion technology for terrestrial applications already exists, much of the first usage was in the US space program. But as the nation's need for electrical energy increases and certain technological advances reduce the cost of photovoltaics, we will begin to see expansion of their use in terrestrial applications.

We have shown that these solar cells do exist and that they actually convert the sun's energy into electricity. Now all we have to do is select the best technologies and proper production techniques in order to reduce their cost sufficiently to make them competitive with other sources of electricity. But because of the limited demand

Fig. 9-4. Photovoltaic array on commercial testing facility (courtesy Motorola).

and the high cost of the photovoltaic arrays, they will not play a significant role in the electrical generation field until more research and development can be done to reduce the cost of photovoltaic systems. The federal government has undertaken to do just that by funding many large-scale photovoltaic applications to try to reduce the price of the cells and to increase the demand. It is unlikely that good results could be achieved in the near future without federal assistance.

A major effort in the photovoltaic program is currently being directed to commercial applications of concentrator systems and single-crystal-silicon flat-plate arrays. A wide variety of other photovoltaic materials now exists and has potential for providing solar arrays at significantly lower prices in the future. Because the use of silicon cells is a proven technology, it is the logical choice with which to begin. However, much research, development, and money will be required to reach these goals in time to meet the energy demands. Solar energy must have time to grow to meet these demands. Solar energy is here to stay, and in time, it can and will provide the needed energy directly from the sun.

Glossary: Solar Electric Terms

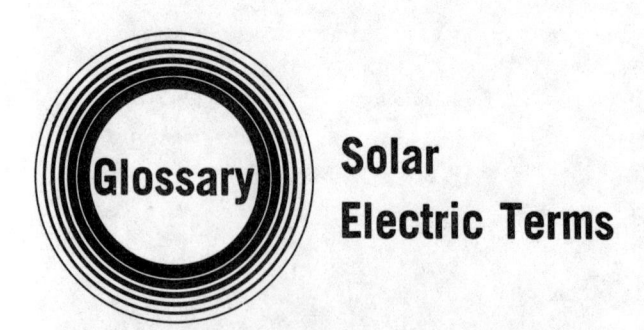

absorber—The surface in a collector that absorbs solar radiation and converts it to heat energy; generally, matte black surfaces are good absorbers and emitters of thermal radiation while white and metallic surfaces are not.

absorptance—The ratio of energy absorbed by a surface to the energy striking it.

AC generator—Producer of electricity using slip rings and brushes to connect armature to external circuit. Output is alternating current.

active system—A solar heating or cooling system that requires external mechanical power to move the collected heat.

air-type collector—A solar collector that uses air as the heat transfer fluid.

alternating current—Current of electrons that moves first in one direction and then in the other; abbreviated AC.

altitude—The angular distance from the horizon to the sun.

ampere—Electron or current flow representing the flow of one coulomb per second past a given point in a circuit.

ampere-hour—Capacity rating measurement of batteries. A 100 ampere-hour battery will produce, theoretically, 100 amperes for one hour.

angstrom—A unit of length used in the measurement of light. 1 angstrom equals 10^{-8} centimeter.

armature—Revolving part in generator or motor. Vibrating or moving part of relay or buzzer.

atom—Smallest particle that makes up a type of material called an element.

atomic number—Number of protons in the nucleus of a given atom.

atomic weight—Mass of the nucleus of an atom in reference to oxygen, which has a weight of 16.

auxiliary heat—The extra heat provided by a conventional heating system for periods of cloudiness or intense cold, when a solar heating system cannot provide enough heat.

azimuth—The angular distance between true south and the point on the horizon directly below the sun.

band—Group of adjacent frequencies in the frequency spectrum.

battery—Several voltaic cells connected in series or parallel. Usually contained in one case.

Btu or British thermal unit—The quantity of heat needed to raise the temperature of 1 pound of water 1° F or to raise the temperature of 0.45 kilogram of water 0.56° C.

by products—Something that is produced in the manufacturing process, other than the main product, that has a value of its own.

calorie—The quantity of heat needed to raise the temperature of 1 gram of water 1° C.

capacitance, distributed—The capacitance in a circuit resulting from adjacent turns on coils, parallel leads and connections.

capacitor—Device which possesses capacitance. Simple capacitor consists of two metal plates separated by an insulator.

carrier—Conducting hole or electron.

circuit breaker—Safety device which automatically opens a circuit if overloaded.

circuit, etched—The method of circuit board production in which the actual conduction paths on a copper-clad insulation board are coated with an acid resist, the board placed in an acid bath and unprotected parts of the copper clad are eaten away, leaving the circuit conductors. Components are mounted and soldered between the conductors to form the completed circuit.

circuit, printed—The method of printing circuit conductors on an insulated base. Component parts may also be printed or actual components soldered in place.

circulating current—Inductive and capacitative currents flowing around parallel circuit.

collector—Any of a wide variety of devices (flat-plate, concentrating, vacuum tube, greenhouse, etc.) used to collect solar energy and convert it to heat.

collector efficiency—The ratio of heat energy extracted from a collector to the quantity of solar energy striking the cover, expressed in percent.

collector tilt—The angle between the horizontal plane and the solar collector plane.

concentrating collector—A device that uses reflective surfaces to concentrate the sun's rays onto a small area, where they are absorbed and converted to heat energy.

conductance—Ability of a circuit to conduct current. It is equal to amperes per volt and is measured in mhos.

$$G = 1/R$$

conduction—The transfer of heat energy through a material by the motion of adjacent atoms and molecules.

conduction band—Outermost energy level of an atom.

conductivity, N type—Conduction by electrons in N-type crystal.

conductivity, P type—Conduction by holes in a P-type crystal.

conductor—Material which permits the free motion of a large number of electrons.

convection—The transfer of heat energy from one location to another by the motion of fluids that carry the heat.

coulomb—Quantity representing 6.28×10^{18} electrons.

covalent bond—Atoms joined together, sharing each other's electrons to form stable molecule.

cover plate—A sheet of glass or transparent plastic that sits above the absorber in a flat-plate collector.

crystal—A substance which has a high degree of order in the arrangement of its atoms. The atoms are usually arranged in a definite repeating pattern.

crucible—A pot or vessel, usually of earthenware, to which great heat can be applied; used in melting metal, ore, etc.

DC generator—Producer of electricity with connections to armature through a commutator. Output is direct current.

degree-day—A unit that represents a 1° F deviation from some fixed reference point (usually 65° F) in the mean daily outdoor temperature.

design heating load—The total heat loss from a house under the most severe winter conditions likely to occur; a concept used in the design of buildings and their heating systems.

diffuse radiation—Sunlight that is scattered from air molecules, dust and water vapor and comes from the entire sky vault.

diode—A two-terminal electronic component that conducts electricity more easily in one direction than in the other.

direct current—Flow of electrons in one direction, abbreviated DC.

direct methods—Techniques of solar heating in which sunlight enters a house through the windows and is absorbed inside.

direct radiation—Solar radiation that comes straight from the sun, casting shadows on a clear day.

doping—Adding impurities to semiconductor material.

double-glazed—Covered by two panes of glass or other transparent material.

dry cell—Nonliquid cell, which is composed of a zinc case, carbon positive electrode and a paste of ground carbon, manganese dioxide and ammonium chloride as electrolyte.

efficiency—Ratio between output power and input power.

electromotive force—Force that causes free electrons to move in conductor. Unit of measurement is the volt.

electron—The smallest division of negative electricity, equal to 1.59×10^{-19} coulombs. Negatively charged particle.

electron volt—A measure of energy. It represents the energy acquired by an electron while passing through a potential of one volt.

element—A substance containing atoms, all of the same kind of material, which cannot be broken down into other substances by chemical means.

emission, types of—Photoelectric is the emission of electrons as result of light striking the surface of certain materials.

emission: thermionic—Process where heat produces energy for the release of electrons from the surface of the emitter.

energy—Material power of the universe; ability to do work; mental or physical force.

energy gap—The energy range between the valence band and the conduction band in a semiconductor.

ERDA—Energy Research and Development Administration.

eutectic salts—A group of materials that melt at low temperatures, absorbing large quantities of heat and then, as they recrystallize, release that heat. One method used for storing energy as heat.

flat-plate collector—A solar collection device in which sunlight is converted to heat on a plane surface, usually made of metal or plastic. A heat transfer fluid is circulated through the collector to transport heat for direct usage or storage.

forbidden region—A region between the valence and conduction band of an atom.

forced convection—The transfer of heat by the flow of warm fluids, driven by fans, blowers or pumps.

fuse—Safety protective device which opens an electric circuit if overloaded. Current above the fuse rating will melt fusible link and open circuit.

galvanic corrosion—The condition caused by a conducting liquid making contact with two different metals which are not properly isolated physically and/or electrically.

generator—A source of electrical energy; converts one form of energy to electric energy.

germanium—A rare grayish-white metallic chemical element.
grid—Grid of fine wire placed between cathode and plate of an electron tube.

header—The pipe that runs across the top or bottom of an absorber plate, gathering or distributing the heat transfer fluid from or to the grid of pipes that run across the absorber surface.
heat exchanger—A device, such as a coiled copper tube immersed in a tank of water, that is used to transfer heat from one fluid to another through a separating wall.
heat pump—A mechanical device that transfers heat from one medium at lower temperature (called the heat source) to another medium at higher temperature (the heat sink), thereby cooling the first and warming the second.
heat sink—A medium to which heat is added (see heat pump).
heat source—A medium from which heat is extracted (see heat pump).
heat storage—A device or medium that absorbs collected solar heat and stores it for use during periods of inclement or cold weather.
heliostat—A mirror used to reflect the sun's rays into an absorber.
henry—Unit of measurement of inductance. A coil has one henry of inductance if an emf of one volt is induced when current through inductor is changing at rate of one ampere per second.
hole—Positive charge. A space left by a removed electron.
hole injection—Creation of holes in semiconductor material by removing electrons with a strong electric field around point contact.
horsepower—33,000 ft-lbs of work per minute or 550 ft-lbs of work per second equals 1 horsepower. Also, 746 watts = 1 hp.
hybrid solar energy system—A system that uses both active and passive methods in its operation (see active system and passive system).

impurity—Atoms with a crystalline solid which are foreign to the crystal.
indirect system—A solar heating or cooling system in which the solar heat is collected outside the building and transferred inside using ducts or piping and, usually, fans or pumps.
infrared radiation—Electromagnetic radiation from the sun that has wavelengths slightly longer than visible light.
insolation—The total amount of solar radiation—direct, diffused, and reflected—striking a surface exposed to the sky.
insulation—A material with a high resistance or R-value that is used to retard heat flow.
insulators—Substances containing very few free electrons and requiring large amounts of energy to break electrons loose from influence of nucleus.
integrated system—A solar heating or cooling system in which the solar heat is absorbed in the walls or roof of a dwelling and flows to the rooms without the aid of complex piping or ducts.
internal resistance—Refers to internal resistance of source of voltage or emf. A battery or generator has internal resistance which may be represented as a resistor in series with source.

joule—Unit of energy equal to one watt-second.
junction diode—PN junction having unidirectional current characteristics.

kilowatt—A measure of power equal to one thousand watts, approximately 1 1/3 horsepower, usually applied to electricity.
kilowatt-hour—The amount of energy equivalent to 1 kilowatt of power being used for one hour; 3413 Btu, abbreviated kW-h. Common unit of measurement of electrical usage.

langley—A measure of solar radiation; equal to one calorie per square centimeter.

lead acid cell—Secondary cell which uses lead peroxide and sponge lead for plates, and sulfuric acid and water for electrolyte.
life cycle costing—An estimating method in which such long-term costs as energy consumption, maintenance and repair can be included in the comparison of several system alternatives.
light meter—A photographic device used to measure the amount of light.
liquid-type collector—A collector with a liquid as the heat transfer fluid.
load—Resistance connected across circuit which determines current flow and energy used.
loading a circuit—Effect of connecting voltmeter across circuit. Meter will draw current and effective resistance of circuit is lowered.

milli—Prefix meaning one thousandth of.
milliammeter—Meter which measures in milliampere range.
minor carrier—Conduction through semiconductor opposite to major carrier.

network—Two or more components connected in either series or parallel.
neutron—Particle which is electrically neutral.
nickel-cadmium cell—Alkaline cell with paste electrolyte hermetically sealed. Used in aircraft.
nocturnal cooling—The cooling of a building or heat storage device by the radiation of excess heat into the night sky.
nondepleting—Ever renewing.
nonpolluting—The ability to keep from making things unclean or impure.
nucleus—Core of the atom.

ohm—Unit of measurement of resistance.
ohm's law—Mathematical relationship between current, voltage and resistance discovered by George Simon Ohm.

$$I = \frac{E}{R}$$
$$E = IR$$
$$R = \frac{E}{I}$$

one sun—A term used in referring to the power of sunlight falling on the earth. It is equal to 1000 watts of radiation per square meter and is the approximate radiation in clear weather supplied by the sun to surfaces perpendicular to the sun's rays.
overcharge—To charge too much, too great a charge; to overload; put too great a strain on; an overcharge on an electric circuit.

passive system—A solar heating or cooling system that uses no external mechanical power to move the collected solar heat.
peak power rating—The amount of power that will be generated in direct sunlight in the middle of a bright day at 77° F.
percentage of possible sunshine—The percentage of daytime hours during which there is enough direct solar radiation to cast a shadow.
phase—A quantity that specifies a particular stage of progress in any recurring operation, such as an alternating current. Phase is usually expressed as an angle or part of a cycle, in which the complete cycle of operation is equal to 360°. When two alternating quantities pass through corresponding zero at the same time, they are said to be in phase.
photo—Produced by light.
photodiode—A PN junction diode which conducts upon exposure to light.
photoelectric cell—A cell which produces an electric potential when exposed to light (see photovoltaic cell).

photoelectrons—Electrons emitted as a result of light.
photon—A discrete quantity of electromagnetic energy; a quantum.
photosensitive—Characteristic of material which emits electrons from its surface when energized by light.
photosensors—Special solar cells used in areas such as tape or card readers, light meters, spectrum detectors.
photovoltaic—The generation of a voltage at the junction of two materials when exposed to light.
photovoltaic cells—Semiconductor (solid-state) devices that convert solar energy directly into electricity without any moving parts (see photoelectric cell).
PN junction—The line of separation between N-type and P-type semiconductor materials.
point contact—A pressure contact between a semiconductor body and a metallic point.
point contact diode—Diode consisting of point and a semiconductor crystal.
power—Rate of doing work. In DC circuits, $P = I \times E$.
primary cell—Cell that cannot be recharged.
pyranometer—An instrument for measuring solar radietion.

quanta—Definite amount of energy required to move an electron to a higher energy level.

radiant panels—Panels with integral passages for the flow of warm fluids, either air or liquids. Heat from the fluid is conducted through the metal and transferred to the rooms by thermal radiation.
radiation—The flow of energy across open space via electromagnetic waves, such as visible light.
refrigerant—A liquid such as freon that is used in cooling devices to absorb heat from surrounding air or liquids as it evaporates.
reradiation—Radiation resulting from the emission of previously absorbed radiation.
resistance—Quality of an electric circuit that opposes flow of current through it.
resistivity—The property of an electrical conductor that determines the current (amperes) between two points of different potential (volts). The ratio of volts to amperes in a conductor.
retrofitting—The application of a solar heating or cooling system to an existing building.
risers—The flow channels or pipes that distribute the heat transfer liquid across the face of an absorber.

schematic—Diagram of electronic circuit showing electrical connections and identification of various components.
secondary cell—Cell that can be recharged by reversing chemical action with electric current.
selective surface—An absorber coating that absorbs most of the sunlight hitting it, but emits very little thermal radiation.
semiconductor—Conductor with resistivity somewhere in the range between conductors and insulators.
semiconductor, N type—Semiconductor which uses electrons as a major carrier.
semiconductor, P type—Semiconductor which uses holes as a major carrier.
series circuit—Circuit which contains only one possible path for electrons through circuit.
series parallel—Groups of series cells with output terminals connected in parallel.
short circuit—Direct connection across source which provides zero resistance path for current.

shunt—To connect across or parallel with a circuit or a component.
solar cells—See photovoltaic cells.
solar constant—The average intensity of solar radiation reaching the earth outside the atmosphere, amounting to 2 langleys or 1.94 gram-calories per square centimeter; equal to 442.4 Btu/hr/ft^2, or 1395 watts/m^2.
solar house—A dwelling that obtains a large part, but not necessarily all, of its heat from the sun.
solar rights—A legal issue concerning the right of access to sunlight.
solar radiation—Electromagnetic radiation emitted by the sun.
specific heat—The quantity of heat, in Btu, needed to raise the temperature of 1 pound of material 1° F.
spectrum—A continuous range of frequencies; for example, the light spectrum.
spectrum detectors—A device used to separate into a spectrum the lights from distant stars to determine their elements.
static electricity—Electricity at rest as opposed to electric current.
storage battery—Common name for lead-acid battery used in automotive equipment.
sun path diagram—A circular projection of the sky vault, similar to a map, that can be used to determine solar positions and to calculate shading.
switch—Device for directing or controlling current flow in circuit.

thermal capacity—The quantity of heat needed to warm a collector to its operating temperature.
thermal radiation—Electromagnetic radiation emitted by a warm body.
thermistor—Semiconductor device which changes resistivity with a change in temperature.
thermoelectric—When two dissimilar wires are heated, a small voltage is produced.
tilt angle—The angle that a flat collector surface forms with the horizontal plane.
transistor—Semiconductor device derived from two words, "transfer" and "resistor."
trickle-type collector—A collector in which the heat transfer liquid flows down open channels in the front face of the absorber.
trickle charge—When a small amount of current is used to recharge a battery.
tube-in-plate absorber—An aluminum or copper sheet metal absorber plate in which the heat transfer fluid flows through passages formed in the plate itself.
tube-type collector—A collector in which the heat transfer liquid flows through metal tubes that are wired, soldered or clamped to the absorber plate.

ultraviolet radiation—Electromagnetic radiation with wavelengths slightly shorter than visible light.
unglazed collector—A collector with no transparent cover plate.

valence band—The energy level of an atom closest to the nucleus.
volt—Unit of measurement of electromotive force or potential difference.
voltage drop—Voltage measured across resistor. Voltage drop is equal to product of current times resistance in ohms. $E = IR$
voltaic cell—Cell produced by suspending two dissimilar elements in an acid solution. Potential difference is developed by chemical action.
voltmeter—An instrument for measuring voltage.

watt—Unit of measurement of power.
watt-hour—Unit of energy measurement, equal to 1 watt per hour.

work—When a force moves through a distance: measured in foot-pounds and work = force × distance.

X-axis—The horizontal axis of a graph.

Y-axis—The vertical axis of a graph.

Suppliers and Dealers of Photovoltaic Cells and Products

Appendix

ARCO Solar,
20554 Plummer Street,
Chatsworth, CA 91311.

Clover Solar Corporation,
600 E. Colorado Street,
Glendale, CA 91205.
Columbia Chase Solar Energy Division,
55 High Street,
Holbrook, MA 02343.
Comspace Corporation,
350 Great Neck Road,
Farmingdale, NY 11735.

Energy Conversion Devices, Inc.,
1675 West Maple Road,
Troy, MI 48084.

Farwest Corrosion Control Company,
17311 South Main Street,
Gardena, CA 90248.

Innotech Corporation,
Norwalk, CT 06856.
International Rectifier,
233 Kansas Street,
El Segundo, CA 90245.

McGraw Edision Company,
Power Systems Division,
P. O. Box 28,
Bloomfield, NJ 07003.
Motorola,
5005 East McDowell Road,
Phoenix, AZ 85008.
M-7 International Company,
210 Campus Drive,
Arlington Heights, IL 60004.

OCLI,
P.O. Box 1212,
City of Industry, MA 01940.

Pennwalt Corporation,
Automatic Power Division,
Hutchinson Street,
Houston, TX 77002
Poly Paks,
P. O. Box 942,
South Lynnfield, MA 01940.

Sensor Technology, Inc.,
21012 Lassen Street,
Chatsworth, CA 91311.

SES, Inc.,
Tralee Industrial Park,
Newark, DE 19711.
Silicon Material, Inc.,
341 Moffet Boulevard,
Mountain View, CA 94043.
Silicon Sensors, Inc.,
Highway 18 East,
Dodgeville, WI 53533.
Siltec Corporation,
3717 Haven Avenue,
Menlo Park, CA 94025.
Solarama,
220 North Richards Street,
Colorado City, AZ 86021.
Solar Energy Company,
810 18th Street,
Washington, D.C. 20006.
Solarex,
1355 Piccard Drive,
Rockville, MD 20850.
Solar Power Corporation,
20 Cabot Road,
Woburn, MA 01801.
Solar Systems, Inc.,
8100 Central Park,
Skokie, IL 60076.
Solec International, Inc.,
Two Century Plaza, Suite 484,
2049 Century Park East,
Los Angeles, CA 90067.

Sollos, Inc.,
2231 Carmelina Avenue,
Los Angeles, CA 90064.
Spectrolab, Inc.,
12484 Gladstone Avenue,
Sylmar, CA 91342.
Spire Corporation,
Patriots Park,
Bedford, MA 01730.
Sun Trac Corporation,
1674 South Wolf Road,
Wheeling, IL 60090.

Textron, Spectrolab Division,
40 Westminister Street,
Providence, RI 02903.
Tyco Laboratories, Inc.,
Hickory Drive,
Waltham, MA 02154.

Vactec, Inc.,
2423 Northline Ind. Boulevard,
Maryland Heights, MO 63043.

Wm. Lamb Company,
P. O. Box 4185,
North Hollywood, CA 91605.

Zurn Industries, Inc.,
West Eighth Street,
Erie, PA 16501.

Additional Reading

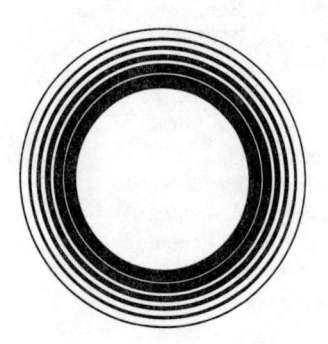

Adams, Robert W., *Adding Solar Heat To Your Home*, TAB BOOKS, No. 1196. Sept. 1979
ARCO, "The Sunny Side of Energy," 1976.
Arizona Solar Energy Research Commission, *Putting the Sun to Work in Arizona*, March, 1976.

Energy Research & Development Administration 77-47/3, *Photovoltaics*, July, 1978.

Foster, William F., *Built-It Book of Solar Heating Projects*, TAB BOOKS, No. 1006. Oct. 1977
Foster, William F., *Homeowner's Guide to Solar Heating & Cooling*, TAB BOOKS, No. 906. Sept. 1976

Gerrish, Howard H., *Transistor Electronics*, Goodhart-Willcos Co., Inc., 1969.

Kaplow, Roy, "A Multiple-Junction Cell for Solar Concentration," *Solar Age Magazine*, June, 1978.
Kuecken, John A., *How To Make Home Electricity From Wind, Water and Sunshine*, TAB BOOKS, No. 1128. July 1979

Merrigan, Joseph A., *Sunlight to Electricity*, M.I.T. Press, 1975.
Mississippi County Community College, "Energy from Solar Cells."
Motorola, Inc., Government Electronics Division, Sales Literature, 1978.

National Solar Heating & Cooling Information Center, "Solar Terminology."

Oddo, Sandra, "Photovoltaics," *Solar Age Magazine*, June, 1978.

Pulfrey, David L., *Photovoltaic Power Generation*, Van Nostrand Reinhold, 1978.

Optical Coating Laboratory, Inc., Photoelectronics Division, Sales Literature.

Sensor Technology, Inc., "An Update on Concentrator Cell Technology and Compatible Concentrator Systems," January, 1978.
Sensor Technology, Inc., "Harvest the Sun," Catalog 250.
Silicon Sensors sales literature, "Technical Bulletins," May, 1978.
Solar Power Corporation (EXXON), "Solar Electric Generator Systems."
Solarex Corporation, "Building Blocks for Solar Electricity," 1977.
Solarex Corporation, "Solar Electricity," November, 1977.
Solarex Corporation, *Making and Using Electricity From the Sun,* TAB BOOKS, No. 1118. April 1979

The Institute of Energy Conversion, *Recent Publications on Photovoltaics*, July, 1978.
Turner, Rufus P., *Solar Cells and Photocells*, Howard W. Sams & Co., Inc., 1977.

United States Department of Energy, "Solar Electricity from Photovoltaic Conversion," March, 1978.
United States Government Printing Office, *Solar Power from Satellites*, January 19 and 21, 1976.

Index

A
Ambient temperature	101
Amperage	38
Antireflective coating	30
Artificial lights	34
Automation	115

B
Band-gaps	59
Batteries	98
nickel-cadmium	102
selecting	98
silver-oxide	102
Battery	34
dry-cell	34
voltage regulators	52
Blocking diode	52

C
Cadmium sulfide	121, 131
Calculators	83
Capsule	44
Cathode protection	76
Cell	37
efficiency improvement	114-124
film	64
high cost	60
how it works	30
made by hand	61
NP type	111
PN type	111
shapes	118
sizes	37
volume low	61
Coal	13
Collector	20
Computer controller	85
Concentrators	128
by-products	138
characteristics	130
collectors	21
materials	131
mounting	132
types	130
Construction equipment	82
Contact points	44
protection	104
Corrosive substances	46
Crystal	26
Current	43

D
Decentralized power applications	68
Deicing chemicals	76
Diesel generators	84
Diode	52
blocking	52
isolation	52
placement	52
Direct current	56
Doping	27
Dry-cell battery	34

E
Electric	82
cars	82
gate openers	82
Electrical	70
cables	70
conversion, methods	124
Electronic applications	70
Electrons	32, 58
Energy	10
alternative natural	15
demands, present-day	12
geothermal	15
solar	16
source of all	10
thermal	124, 140
Etching process	28

F

Film cells	64
Flux solution	29
Flywheel storage	55
Fossil fuels	20
Freon	133
Fresnel lens	137

G

Gallium arsenide	131
Generating systems, small-motor	71
Geothermal energy	15
Grid design	117
Growing ribbons of silicon	117

H

Highway emergency boxes	84
Holes	28, 33
Hydroelectric power	13, 16

I

Indium	121
Ingots	26, 114
Insulators	98
wall	33
Integrated circuit	62
Isolation diode	52

L

Langleys	97
Large scale applications	84
Leads	33
Line power	67
Lithium sulfide	55

M

Marine applications	76
Materials, new	120
Metal grid	29
Metallic coating	33
Metallurgical-grade silicon	26
Microgenerators	38
Mirrors	137
Mississippi County Community College, large solar project	85
Module	30, 42
Molten liquid	114
Multiple cell	119

N

Natural gas	13
Navigational aids	74
Neutrons	32
Nuclear	13
fushion	16
power	13

O

Ocean thermal-energy conversion	148
waves	15
Oil	13
Operating temperatures	101
Optimum design	97
Orange dwarf	10

P

Partial air-conditioning	87
Photocell	58
industry, future goals	62
Photographic process	28
Photophone	72
Photosensors	111
Photosphere	10
Photovoltaic	67, 140
applications	71
cell	22, 150
cell, discovery	22
conversion	152
effect	30
government involvement	145
looking ahead	149
modules replace windmill	85
program plan	65, 153
systems	104, 152
Power	13
conditioner	56
conditioning	98
converters	98
hydroelectric	13, 16
line	67
nuclear	13
output	96
plant	146
requirements	94
wind	15
Primary batteries	71
Protons	32, 58
Pumped storage	55

R

Radiation	92

167

Radio	72
communication systems	72
Railroad crossing	78
Recreational uses	82

S

Sand	26
Seasonal variations	91
Sea water	46
Selenium	22
Semiconductor	58
grade silicon	60
materials	58
Silicon	23, 58, 131
atom	32
cells, how made	26
growing ribbons of	117
metallurgical-grade	26
raw	61
semiconductor grade	60
Sizing the system	97
Small scale applications	72
Solar	16
-driven air-conditioning system	87
-Electric-Application program	148
electric power systems	71
energy	16
generating plants	80
modules	70
One	24, 122
panels	78, 101
photovoltaic conversion	148
thermal conversion	21
thermal-electric conversion	148
Solar arrays	48
cleaning	48
design data, collecting	91
large	48
maintenance-free	48
battery	21, 34
charger	50
Solar cell	22
charging system	52
efficiency	58
life expectancy	102
panels for irrigation	85
Solar-powered station	140
stations	68
Soldering process	42
Sovereign of the Seas	12
Space cells	109
program	68
Storage	50
systems	50
battery	54, 98
cost	54
improvement	54
Sun's angle	94
the forgotten	11
Sunlight, diffused	96

T

Technology, current	113
Terrestrial cells	110
Thermal	16
electrics	16
energy	124, 140
Thermocouples	124
Thermoelectric generators, fossil-fuel fired	71
Thin films	120
Timer	134
Toys	82
Tracking devices	100, 132
systems	135
Transistor industry	61
TV, Educational	76

W

Waste reduction	116
Watches	83
Watts of power	40
Weather	92
conditions	96
patterns	92
Web dendrite	117
Wind-energy conversion	148
power	15
Wood	15

V

Voltage	42
output	42
regulator	52, 98
V-shaped photovoltaic cell arrangement	133

Z

Zinc chloride	55

DISCARDED

JUN 2 4 2025